Decoding Racial Ideology in Genomics

Decoding Racial Ideology in Genomics

Johnny E. Williams
Foreword by Joseph L. Graves Jr.

LEXINGTON BOOKS
Lanham • Boulder • New York • London

Published by Lexington Books
An imprint of The Rowman & Littlefield Publishing Group, Inc.
4501 Forbes Boulevard, Suite 200, Lanham, Maryland 20706
www.rowman.com

Unit A, Whitacre Mews, 26-34 Stannary Street, London SE11 4AB

British Library Cataloguing in Publication Information Available

Library of Congress Cataloging-in-Publication Data
The hardback edition of this book was previously cataloged by the Library of Congress as follows:
Names: Williams, Johnny E., author.
Title: Decoding racial ideology in genomics / by Johnny E. Williams ; foreword by Joseph L. Graves Jr.
Description: Lanham : Lexington Books, 2016. | Includes bibliographical references and index.
Identifiers: LCCN 2016011889
Subjects: LCSH: Race. | Human genome. | Genomics--Social aspects. | BISAC: SOCIAL SCIENCE / Minority Studies. | SCIENCE / Life Sciences / Genetics & Genomics. | SOCIAL SCIENCE / Discrimination & Race Relations.
Classification: LCC GN269 .W55 2016 | DDC 305.8--dc23 LC record available at https://lccn.loc.gov/2016011889

ISBN: 9780739148952 (cloth : alk. paper)
ISBN: 9780739148969 (pbk. : alk. paper)
ISBN: 9780739148976 (Electronic)

∞™ The paper used in this publication meets the minimum requirements of American National Standard for Information Sciences Permanence of Paper for Printed Library Materials, ANSI/NISO Z39.48-1992.

Printed in the United States of America

For my mother, Inez Williams
and in loving memory of my father, Raymond Williams, Sr., and brother,
Byron K. Williams

And for the joyous memory of my best friends and fellow truth seekers
Jerry G. Watts and Lawrence Holcomb

To all those who came before, and to all those who will follow

Table of Contents

Foreword

W. E. B. Du Bois noted that the challenge of the twentieth century would be "the color line."[1] These prophetic words ring true even as we struggle through the beginning of the twenty-first century. Still we observe institutional structures that predominantly serve persons of European descent while discriminating against persons of African descent (as well as other non-Europeans). The most heinous of these the epidemic of police murders against unarmed African Americans. The violence against African American youth partially results from the unconscious view of European Americans that African Americans are not fully human.[2]

Yet, few people know that when Du Bois penned these words, the scientific discoveries capable of dismantling the notion that biological races existed in the human species had already been made. Charles Darwin in *The Descent of Man and Selection in Relation to Sex* originally published in 1871 thoroughly debunked the pseudoscientific beliefs of the polygenists of his day who believed that the races of humankind were in fact separate species. The polygenists thought that these species were, of course, hierarchically ranked with Europeans occupying the pinnacle of racial perfection. While Darwin displayed many of the social prejudices common of the upper social classes of England, he was actively opposed to slavery and unlike many of his fellow Englishmen had spent a great deal of time surveying the social relations of humans in various parts of the world. Most of his observations concerning human biological variation were subsequently shown to be correct and one of the most important of these was his statement that humans were protean or polymorphic. This meant that for the most part the variations displayed in our species were "of an indifferent nature." Thus Darwin reasoned that this variation persisted because natural selection was simply unconcerned with them, meaning that these physical differences had little impact on differential reproductive success.

On this last point, Darwin was partially wrong. This occurred because in 1871 he did not have a fully developed understanding of the mechanisms of heredity. This would be developed over the early part of the twentieth century in what is known as the "Neo-Darwinian Synthesis" which brought together the forces of natural selection and the mechanisms of Mendelian genetics. As genetic studies of human beings progressed across the century, more and more of the notion that legitimate biological races existed in the human species began to erode. Prior to

World War II, anthropologists, geneticists, and evolutionary biologists such as Ashley Montagu, Ruth Benedict, Leslie Dunn, and Theodosius Dobzhansky had recognized the fallacy of the biological race concept in humans.[3] After the war, the world had observed the horrors unleashed by Nazi genocide based upon false conceptions of race. Montagu, Dobzhansky, and others published the United Nations Educational Scientific and Cultural Organization (UNESCO) statements on race beginning in 1950.[4] These pointed out that unambiguous classification of human populations into biological races was not possible, and that we should not expect there to be genetically based differences in complex behavior such as intelligence or moral character between human groups.

Despite the fact that experts in anthropology, genetics, and evolution had made these pronouncements, racism in the United States and around the world progressed unabated after 1950. The scientific data relied upon by the scientists of this period were somewhat crude, but our understanding of human genetic variation would greatly accelerate in the decades that followed with the developments of protein gel-electrophoresis and eventually methods to sequence DNA, now culminating in Next Generation Sequencing (NGS). As this technology developed a core result was validated over and over again. In 1969, Richard Lewontin (Harvard University) found that the genetic variation within "so-called" biological races of human beings was always greater than the genetic variation between them.[5] This meant that there was no legitimate statistical way to define biological races in the human species at the genetic level. Some dismissed Lewontin's reasoning because he was known to have "leftist" tendencies, but other population geneticists without such reputations also found these same results.[6] Since this time, all studies of human genetic variation, whether by restriction fragment length polymorphisms (RFLPs) or the most modern NGS (single nucleotide polymorphisms (SNPs) and structural variants (CNVs) have shown the same results. Thus modern anthropology, genetics, and evolution has clearly demonstrated that biological races do not exist in the human species.[7] This view was recently reiterated by eminent genomic researchers and sociologists in *Science*, the publication of the American Association for the Advancement of Science (AAAS).[8]

Given the narrative above, the reader may now be wondering why this book is necessary. While I would argue that the best minds in my field understand the fact of the nonexistence of biological races (e.g., Francis Collins's statement in *Nature Genetics* 2004), this message is still not widely appreciated.[9] I can speak to this fact personally, having been in attendance at the conference in 2004 that generated Collins's statement, as well as many other conferences since attempting to generate dialog between geneticists and social scientists. Professor Johnny Williams provides us here with a masterful analysis of how many modern genomic scientists engaged in human research continue to utilize protocols that are grounded in implicit racist assumptions and perceptions. Many of these genomic scientists have not been sufficiently trained in or exposed to the long narrative in evolutionary biology concerning the concept of biological race. Thus, as I have pointed out many times in the past, they assume that the socially defined races of U.S. society are legitimate biological entities or that they can be useful proxies for underlying genetic variation.[10] In this work Professor Williams lays bare the role of white supremacist ideology in distorting crucial genomic research that has profound implications for human health and well-being. The book should be read by anyone engaged in human genomics research, as well as anyone who has an interest in social justice.

Dr. Joseph L. Graves Jr.
Associate Dean for Research and Professor of Biological Sciences, Joint School of Nanoscience and Nanoengineering
North Carolina A&T State University and University of North Carolina, Greensboro
Fellow, American Association for the Advancement of Science, Section G: Biological Sciences

NOTES

1. Du Bois, *The Souls of Black Folk*.
2. Eberhardt et al., "Seeing Black: Race, Crime, and Visual Processing."
3. For example, Ashley Montagu's *Man's Most Dangerous Myth: The Fallacy of Race* was first published in 1945.
4. Hazard, *Postwar Anti-Racism: The United States, UNESCO, and "Race," 1945–1968.*
5. Lewontin, "The Apportionment of Human Diversity."
6. Nei and Roychoudhury, "Genic Variation Within and Between the Three Major Races of Man: Caucasoids, Negroids, and Mongoloids."
7. Graves, "Why the Nonexistence of Biological Races Does Not Mean the Nonexistence of Racism."
8. Yudell et al., "Taking Race out of Human Genetics."
9. Collins, "What We Do and Don't Know about 'Race', 'Ethnicity', Genetics and Health at the Dawn of the Genome Era."
10. Graves, "Evolutionary versus Racial Medicine."

Preface

Systemic racism is one of the most insidious and violent oppressions humans have ever concocted to legitimate exploiting fellow human beings. Yet, 'white' racial oppressors and their racially marginalized collaborators usually find it difficult to comprehend and acknowledge their complicity in the systemic abuse of racialized 'others.' Often the difficulty with racial oppressors comprehending how racism works in and through their lives arises from their thorough immersion in and internalization of white supremacy narratives and assumptions about whiteness and racialized 'others.' Because of this, historian John Hope Franklin aptly observed:

> The specter of color is apparent even when it goes unmentioned, and it is all too often the unseen force that influences public policy as well as private relationships. There is nothing more remarkable than the ingenuity that the various demarcations of the color line reflect. If only the same creative energy could be used to eradicate the color line; then its days would indeed be numbered.[1]

The endurance of racist perceptions and conventions are fueled in part by the deeply grounded commonsensical view that racism is an individual rather than a structural phenomenon and thus not sustained by 'white' control over institutional tools of power. Racism is a network of interlocking, reinforcing institutions of all organizational domains: political, economic, social, cultural, legal, military, education. As a system, racism affects every aspect of life. So, 'white' disregard of racism as a systemic phenomenon guarantees it continuous operation in society.[2] Given this, I argue in the book that it is essential that genomicists, particularly self-identified 'white' ones, comprehend white supremacy as systems of power relations to ameliorate genomics' endemic racial reasoning and practices.

In genomics racial ideology most often manifests itself through standards and protocols by which it is evaluated as a credible science. These standards and protocols are usually grounded in implicit racist assumptions and perceptions that most genomicists fail or refuse to acknowledge as factors in their research. Though the last few decades have seen an increase in scholars' interests in how power and social location shape human genome knowledge and its application, very few genomicists consider how racial ideology informs their thinking and perceptions of human genome variation. This is so mainly because of the racist cognitive structures inscribed in genomicists' minds and actions, and their adherence to the myth of *objectivity*, which lead most to overlook the simple fact that structures and culture are inseparable. How genomicists practice and think about genetic reality are shaped—though no way determined—by the larger circumstances in which they find themselves. Because most scientists believe their research is independent of sociopolitical influences, they rarely consider how their racial social situatedness informs their research. Professing objectivity or value neutrality, then, discourages the critical reflexivity required for genomicists to account for the racism and racial ideology implicated in their research. This inhibition helps to fuel racist thinking and practices that normalizes whiteness to sustain racially oppressive social systems. To effectively reduce the impact of racist thinking in genomics, researchers need to concede that racial ideology is rooted in genomic methodology and reasoning, and that genomics plays a significant role in promoting racial thinking and racism through its constant search for racial genetic differences.

I wrote this book because I believe that white supremacy as an all-encompassing force threatens the survival of humanity with its distortion of human genetic variation knowledge. For almost a half a millennium now, racism has been the primary source of cultural consciousness. So-called 'whites' from all walks of political, social, and economic life have and continue use to racial ideology to construct reality to comport with their need to justify racial hierarchy. The chief legitimators of white supremacy come largely from the intellectual class working in universities and colleges. Their claim that intellectualism/science is a disinterested, objective pursuit of 'pure' knowledge is profoundly dubious. Scientists are keenly aware of their power to shape the public and the scientific community's thoughts about phenomena through their never-ending effort to publicize their ideas and findings through various media. Such self-promotion in the academy is rampant in ways that tend to serve nefarious political and economic agendas. To ensure that their research resonates with readers/audiences, researchers in their effort to garner funding and public notoriety use familiar racial conceptions of genetic reality. Such schemes are evident in glitzy public relations campaigns undertaken by genomic researchers and publishers. The fact is genetic findings are usually not as simple as public relations firms and media present them. Genetics is complicated and contingent upon the many forces acting on it.

Fortunately, the academy and academic conferences provide a little more rigorous public forum for scientific work that is more multifaceted with a less spectacular presentation. For genomics, unlike commercial media, does not convey the simple objective truth that 'race' is biological. Rather most genomicists declare 'race' has no biological basis. In spite of this assertion people's perceptions of one another as racial beings continues, and it persists in no small part due to these very same genomicists re-inscribing 'race' as minuscule molecular differences between groups produced by environmental variability among them. This illustrates the deeply engrained influence of white supremacy sense making in the minds of most self-identified 'white' genomicists—and even, at least in part, many racially oppressed scientists searching for racial genetic differences. This deeply held cultural consciousness is broad and encompasses many pieces of racialized knowledge and understandings working in concert to shape genomicists' commonsensical action, perceptions, and thinking about human genetic variation.

Given that there are multiple layers to the insidiousness of white supremacy in science and the larger society, I am certain that readers of this manuscript will experience many feelings. Let these emotions guide you through your uneasiness and doldrums of inaction to participate in the mental and physical dismantling of white racism in genomics. For as the late South African human rights activist Steve Biko observed: "white racism is only possible because whites are indifferent to suffering and patient with cruelty" meted out to black folks. Instead of involving themselves in all-out efforts to dismantle white supremacy, 'whites' waste time trying to prove to the racial oppressed that THEY are not their oppressor.[3] The fact is that systemic racism and its ideologies are propagated primarily by 'white' genomicists who predominate in genomics, who need to redirect their focus to ameliorate the influence of their racial conditioning in genomics. Racism's omnipresence requires genomicists not shirk the reality of their complicity in it by retreating into the impotency of despondence and personal guilt. Rather genomicists, particularly self-identified 'white' researchers, should constantly problematize racial thinking in their reasoning and methodologies in order to tackle the menace of racism and racial ideology residing within them and genomics. For the first act of a critical genomicist is to excise racism's inhumane manifestations in genomics. This is the only path to liberating genomics from white supremacy.

NOTES

1. Franklin, *The Color Line: Legacy for the Twenty-first Century*, xiii.
2. Martinez, "What is White Supremacy?"
3. Biko, *I Write What I Like*, 23.

Acknowledgments

I am grateful for the intellectual support, guidance, and encouragement of many people in helping me work through the many years of thinking and writing to complete this study of what most people consider an uncomfortable and difficult subject to grapple with: genetics and white supremacy. Exposing one's private thoughts to intellectual scrutiny can be difficult, especially when those thoughts contradict self-presentation. Yet had it not been for the willingness of genomicists to share their experiences, I could not have written this book. I want to thank you for taking the time away from your demanding schedules to speak with me.

I also extend my deepest gratitude to Tanetta Andersson, Noel Cazenave, Ashley "Woody" Doane, Stephen Valocchi, and Melissa Weiner for their invaluable comments and suggestions on drafts of the book. The editing skills of my former undergraduate student and now Boston University PhD candidate Cara Bowman were an enormous help in making my writing about the complex interaction between racialized culture and genomic knowledge production more intelligible. I also thank Claire Reinelt for critical feedback that pushed me to clearly express my thoughts. Thanks also to the editors at Lexington Press for patiently waiting for me to complete the manuscript.

I am grateful to my deceased friends and colleagues Lawrence Holcomb and Jerry G. Watts for encouraging me to believe in my abilities to retrain in what we all initially thought was an esoteric and difficult to understand new genetic methodology: human genomics. You may not be here physically but you all are always with me mentally and spiritually. Your input for the transformation of the world into a better place for all continues through me and your students.

During the research process, I benefitted from the critical attention of my mother, Inez Williams, who though was never really keen on talking about genetics and genomics engaged me in conversations about the everyday workings of oppression that helped me think about the book's topic in ways that would never had availed themselves to me otherwise. Thank you, Momma! Finally, I want to think my life partner, Elisabeth Neiterman, and our children, Paki (pah-key) and Rachael Williams, for their grating but loving reminders that there is more to life than academic articles and books. Who knew! All kidding aside, I want to thank them for putting up with my impatience, neurotic tendencies, and chaos while writing the book. I love you all very much and very deeply.

Introduction

Although many genome scientists assume that scientific knowledge is free from the effects of social life, scientists' genetic theorizing and thinking emerges in part from the sociohistorical context in which they are immersed. Genome scientists, like everyone else, are a part of society not outside of it. They share the same beliefs and prejudices as others in society and operate within the dominant ideological frames of their specific sociohistorical contexts. Their embeddedness in social formations and conditions with distinct ideologies deeply shapes their views, perceptions, and interpretations of genes. As members of societal groups, genomicists do not make sense of the genome as discrete solitary individuals but as interacting socialized persons equipped with their particular social group's cultural meanings about biological reality.[1] For instance, genomicists who self-identify as 'white' Euro-Americans share experiences and meanings with other 'white' Euro-Americans that directly and indirectly shape their perceptions and understandings of human genome variation.[2] The collective experience of being a 'white' Euro-American activates shared meanings and other cultural processes (e.g., cognitive, emotional, motivational, and practical) that inform how 'white' genomicists discuss and practice genomics. Although genomicists are generally aware that the genome acquires meaning through human definition and practices, most assume the posture of value-free and objective scientists acting as if the sociopolitical world has no bearing on their research.

Because decisions about basic assumptions guiding scientific research are social phenomena, genome research and theory are not solely understandable by simply monitoring empirical progress in the field.[3] In this book, I examine how racism and its associated ideologies impact how genomicists construct and talk about human genome variation knowledge. How genome investigators acquire and make sense of research data is driven by certain social assumptions that are uncritically accepted even in the face of empirical refutation. For example, though it is well established that 'race'[4] is a sociohistorical creation, sociologist Sandra Lee found that most human geneticists routinely racially label genetic material to facilitate sorting them into groups with presumed genetic homogeneity.[5] I examine how practices of this sort go largely unquestioned because they coincide with many genomic researchers' commonsense 'cultural repertoires' about 'race.' Given that 'race' operates as ideological common sense I approach genomic thought, form, and content in the book as partly an outcome of genomicists' racialized worldviews.

Because racial ideology as common sense is a pervasive belief system consisting of language, images, myths, ideas, and values I examine how genomicists use this perceptual frame to give meaning to genetic facts and events.[6] As a commonplace perceptual and ideological frame, 'race' is also used to convey assumptions and expectations about reality while interacting with other features of culture's perceptual structure (e.g., language, symbols, images, etc.). For example, scientists' racial biological representations embody expectations; scientific words contain racial assumptions and symbols representing racial meaning.[7] Genomicists using 'race' as a central perceptual frame use it as a guidepost for determining which facts are important, which to ignore, and which to take for granted.[8] Given this, I consider how genomic explanations and representations are often constructs of collectively accepted cultural cognitive frames about genetic reality and socially established ways of interpreting and understanding the human genome. I approach racial ideology then as an enduring cultural component that does not simply give meaning to superficial physiological differences; it operates as an unexamined and unchallenged way of thinking and belief entrenched in and reproduced continuously by people in every medium and institution in society.

In addition to its societal ubiquity, I highlight the importance of 'race's' ideological elasticity—its ability to persist in the genetic field when challenged. This flexibility is unmistakable among promoters of color-blind ideology. The devotees of color-blind ideology contend racism is no longer a problem since the modern civil rights movement successfully dismantled it in the 1960s, so noticing or considering racial differences indicates one is racist.[9] Since being color-blind—or perhaps better put, racism-evasive—is considered positive, proponents of the ideology working in genomics are hesitant to acknowledge and talk directly about the 'race' concept and systemic racism's impact on their thinking and methodological designs. But their reluctance does not explain why most genomicists generally accept the color-blind narrative as fact. In this book I emphasize how this acceptance is due in part to color-blind ideology's not so subtle shifting of culpability for the persistence of racial inequality away from 'whites' to nonracial dynamics (e.g., malintegrated individuals, abnormal 'black' culture). Such racial indirectness allows 'white' genomicists to benefit from what George Lipsitz calls the "possessive investment in whiteness"—while shoring up their 'anti-racist'[10] bona fides to avoid confronting their culpability in sustaining white supremacy.[11] Similar to sociologist Dorothy Roberts, I argue that a key area in which this materializes is when 'white' genomicists relegate racial bias in genetics to an earlier 'unsophisticated' scientific genetic era.[12] By assigning responsibility for scientific racism to a bygone genetic era, genomicists frame contemporary genomics as an exceptional site of scientific 'race' neutrality.[13] This maneuver exposes color-blind ideology for what it really is, a means for researchers to actively participate in producing and supporting racialized structures and practices without acknowledging that this is in fact what they are doing.

It is impossible to rid genomics[14] of racial thinking without getting rid of the language and discourse concealing it. Because language and discourse are necessary mediations for comprehending reality, I investigate how racial ideology sustains racist thinking in genetics. In this chapter I introduce the problem of racial ideology in genomics and briefly explain its place within the larger racial historical and social context. This discussion is followed by a concise summary of scholars' work on racial meaning in genome science. I end with a review of the theoretical scholarship I find most relevant for the studying racial ideology's effect on human genome analyses.

HISTORICAL CONTEXT OF RACISM AND RACIAL IDEOLOGY IN GENOMICS

The contemporary workings of racial ideology in genomics can be best understood when viewed in their racial historical context. The making of human variation into 'race' began in 1758 with Swedish zoologist and physician Carolus Linnaeus's division of humans as a species into four geographical subgroups or what he termed 'varieties' in *Systema Naturae.*[15] Expanding on Linnaeus's scheme, German professor of medicine and zoology Johann Friedrich Blumenbach in *On the Natural Variety of Mankind* introduced a comparative anatomy race-based classification scheme that increased the original geographically based four-race arrangement to a five-group one. Blumenbach's five categories were: Caucasian, the 'white' race; Mongolian, the yellow race; Malayan, the brown race; Ethiopian, the black race; and American, the red race. The Age of Enlightenment provided the backdrop for 'scientifically' corroborating the existence of 'races' in order to facilitate racism. At this early stage of scientific racism, Charles Darwin came forth with his theory of evolution to problematized 'race.'[16] In his 1871 publication *Descent of Man and Selection in Relation to Sex,* Darwin laid the scientific foundation for thinking critically about the idea of biological 'races' within the human species.[17] In *Descent of Man* Darwin problematized both the notion of separate 'races' and the classification schemes that zoologists like Linnaeus and Blumenbach used to categorize people into 'races.'[18] He argued that the arbitrary way in which physical traits were chosen for inclusion in racial classification schemes were inexact. The vagueness of traits ensured groupings ranged wildly from two to sixty-three different racial groups thus making them scientifically untestable.[19] Even if zoological classifications had agreed, evolutionary biologist Joseph Graves argues they would have failed due to the *principle of discordance* which stipulates physical traits are controlled by different segments of the genome influenced by selection pressures in particular environments.[20] This means that phenotypes such as skin color, hair type, and so forth mix and match in different human groups, making discerning racial relatedness via physical traits impossible.[21]

With evolutionary biology's rise to prominence in the twentieth century, scientists acquired the means to test the validity of biological 'race' claims. Data analyzed by social and natural scientists countered the idea that biological 'races' exist within the human species.[22] The essentialist ideas of varieties and species gave way to Darwin's evolutionary biological reasoning that specified that human groups accumulate genetic variants that are useful for survival and reproduction within specific environments. According to Darwin, if genetic adaptation sustains itself over time, in the absence of mating between different human groups, gene frequencies may diverge resulting in the formation of new human species. However, given that humans maintain sufficiently high intergroup mating rates, reproductive isolation does not necessarily mean that genetic frequency variation within a group signifies bioracial distinction.[23] Darwin's evolutionary biological deductions make it clear that 'races' are a level of genetic divergence that occur somewhere between a single unitary group living in one set of environmental circumstances and the formation of multiple groups adapting to different environmental circumstances.[24] Plainly stated, human "populations"[25] become genetically different because they are changed by the different environments they inhabit.[26] Consequently, the question before contemporary genomicists remains how and when is there enough genetic variation between human groups to classify them as 'races?'[27] In the book, I make the case for the genomic variation field's obsession with establishing the existence of human 'races' as an outcome not of genetic fact but racially conditioned genomic thinking and practices. I reject the notion that genomics transcends the sociopolitical environment in which it operates based on what I consider to be a more accurate, interactive, and critical view immersed in systemic racism. Because systemic racism as a system of concepts and beliefs organizes thought, policies, and practices, my analysis goes beyond merely focusing on how it facilitates 'white' material advantage to consider how racial ideology permeates most genomicists' minds as normative sense-making frames.

TERMS AND DEFINITIONS

Terms like ideology, 'race,' systemic racism, and racialized culture are frequently expressed in both academic and public discourse without providing a precise definition. Given the lack of specificity about these terms, I provide my particular definitions here.

Ideology

According to sociologists Karl Mannheim, no individual "confronts the world and in striving for truth constructs a worldview out of . . . [individual] experience. . . . [Rather] knowledge is from the very beginning a co-operative process of group life, in which everyone unfolds . . . knowledge within a framework of a common fate, a common activity, and the overcoming of common difficulties."[28] Since knowledge—by which he means "evidence-based representations of the natural, social, and behavioral world"—is socially conditioned, it is important to evaluate how scientists' cognitive frameworks and conceptual schemes are influenced by social conditions and their social locations.[29] Often, these cognitive and conceptual frames obscure the world even as they provide frameworks for representing the world.

Mannheim (*Ideology and Utopia: An Introduction To The Sociology Of Knowledge*), and Peter Berger and Thomas Luckmann (*The Social Construction of Reality: A Treatise in the Sociology of Knowledge*), make important and quite different contributions to understanding how social antecedents are made real in knowledge production. Mannheim focuses largely on the role that ideology plays in our representations of the workings of the social world within which we live. Berger and Luckmann focus on the specific ways people acquire "ordinary knowledge" and incorporate commonsensical understandings into their knowledge about reality.[30] These knowledge frames are creations of major historical and cultural work which people add to as they create fresh knowledge. Hence it's important to approach genomics' conceptual frameworks and explanatory hypotheses as contingent and historical productions that have a social history.

Given the contextual contingency of knowledge creation, Louis Althusser's "ideological interpellation" offers an important analytical lens for understanding how genomicists' internalization of racial ideology is insinuated in human genome knowledge production.[31] By ideological interpellation Althusser means the symbolic identity given to us by the reigning rulers of society to make us into subjects who embody the dominant cultural worldview. This concept provides insight into how supposedly value-neutral genomicists accentuate or ignore genome thought incongruent with racial ideology. Specifically, racially interpellated genomicists reflexively construct what philosopher Charles Mills calls "epistemologies of ignorance" that conceal from themselves and the public how genomics contributes to the production of racialized social relations and conformity through the distortion of powerless racial groups' genetic makeup.[32] These biological misrepresentations encourage the public and genomicists to largely accept 'race' and racism as natural, thus beyond modification.

With the ideology concept I consider how genomic thought, form, and content are outcomes of racially subjected genomicists. Specifically, I investigate how racialized social conditioning affects the basic orientation, evaluation, and content of scientists' ideas about the human genome, as well as the manner by which they approach, formulate, and observe genome variation. I do not contend, however, that racialized environments are strict determinates of genomic thinking, but rather genomicists' structural location in racialized societies shape genome thought and practices. In this study I treat racialized societies and their cultures as possessing distinct racial ideologies, not merely bodies of ideas, but lived material practices—ways of thinking taking actual form—which are reproduced through the construction of genome science.

Race and Systemic Racism

Ideally in the social sciences the 'race' concept is supposed to refer to arbitrary opinions about socially and historically defined racial groups' physiology but more often than not social scientists' focus does not go beyond 'race' as a social construction. The contention that 'race' is a social construct confines social sciences' analysis of 'race' to that of an attitude without real material connections. 'Race' is not a physical fact, but a product of racism. Racism is not an attitude or a state of mind, it's an action. In spite of this, 'race' is presented in the social sciences as merely an ideological construct devoid of any link to racism. I do not separate 'race' and racism in this manner, because there can be no 'race' without racism and vice versa. Both are unthinkable without one another. 'Race' is a historically specific ideology that emerged, took shape, and evolved as a constitutive element within a definite set of social relations anchored the capitalist systems of production.[33] As an imagined idea that human beings are ordered in distinct biological varieties, 'race' took shape and evolved in biology to become a typology for subgroups within the human species.[34] These subgroups, which biologists termed subspecies, posit the existence of sharply genetically differentiated geographical groups within the human species.[35] This is how the 'race' concept, as I discuss in chapter 2, was established as a reasonable scientific fact.

Though 'race' is a cultural invention of science rather than biological construct, it is nonetheless real.[36] 'Race' is structured in the rhythms of everyday life as racism. It is lived, concrete. Every part of the life cycle, and most aspects of one's life, is shaped by racism which is integral to the foundation of genome knowledge production.[37] Racism's ideologies shape what genomicists do, how they think and interact with one another and organize their daily lives. 'Race,' as an ideology developed alongside of Western colonialism, is a foundational pillar of Western social structures that was enshrined by nineteenth century natural scientists to legitimate 'whiteness' and the material and psychological advantage of individuals labeled as 'white.' It is important to recognize, however, that 'whiteness' is not about phenotype/genealogy, it is a social marker for allegiance to power relations grounded in systemic racism.[38] Basically, systemic racism as a historically based, institutionalized system of exploitation and oppression functions to maintain 'white' wealth, power, and privilege. Thus racism is not merely a manifestation of individual racial prejudice; it is a system of power relationships justified by the ideology of 'race.' As a form of oppression, racism is used by 'whites' to exercise control over racialized 'others.' Today this power often operates imperceptibly within hierarchical systems of domination—major societal institutions like science—that generate and sustain society's material, social, and ideological reality.[39]

As a system of deeply seated social and institutional practices, systemic racism encourages individuals to perceive human differences not as they exist in the world but through racialized cultural frames that naturalize racial social arrangements. In this sense systemic racism is produced simultaneously at both the cultural and structural level. Racism at the cultural level consists of values, norms, and shared patterns of seeing, thinking, and acting which validate and legitimate a 'white' worldview. This way of knowing and perceiving reality is transmitted and presented in a manner that secures the oppression of subordinate socially defined racial groups. For example, cultural mechanisms such as the media—especially popular culture—transmit the idea that genes determine the hierarchical ranking of socially defined racial groups. Systemic racism's structural reproduction involves institutional actions, decisions, and policies that systematically advantage 'whites' at the expense of other socially defined racial groups.

'Race' as an ideological construct constitutes individuals and groups identifying as 'white' and believers in the ideology of whiteness who defend white supremacy as an exploitive oppressive system of social differences hierarchically organized around social categories of 'race.' Although systemic racism benefits 'whites,' it is more than a collection of 'white' people acting in their own interest. It is an all-encompassing social system that both constrains and shapes all societal members' lives. 'Race' as a central ideology is a part of prevailing forms of consciousness generated within the cultural terrain in which people "move, acquire consciousness of their [social] position, struggle, etc."[40] Therefore as a body of interconnected ideas about how and why society is ordered the way it is, racial ideology configures people's understanding of reality.[41] These racial ideas, however, do not simply emerge out of thin air. They come from racial and class positions and concrete interests that incline people to act in accordance with racial ideology. This means that people's language and actions in racialized cultures are the material expressions of racial ideology; it is racial meaning articulated in practice.[42] Because racial ideology and genomics both emerge out of racist conditions of existence, it is imperative that we deconstruct the relations between the two to determine how genomicists' racial worldviews shape their understanding of genome variation.

Racialized Culture

In this book racialized culture is used to refer to 'race' content (shared, meanings, beliefs, values, symbols,) and the interpretive process that transforms content into racial understandings and practices. With this concept, I examine not only how raced cultural norms, symbols, and practices are used to construct biological differences between groups, but also how the biological construction of 'race' responds to changing sociohistorical circumstances. Similar to Pierre Bourdieu's *habitus,* I use racialized culture to explore how socially acquired "deep structural" racial classificatory and assessment propensities manifest in genomicists' outlooks, thoughts, and conceptions of human genome variation.[43] The application of the habitus concept helps illuminate how racialized culture operates among genomicists as a force of habit delimiting parameters for what is 'thinkable/unthinkable' for individuals within a particular raced habitus or structural location. Consistent with Marxist theory, the habitus concept specifies that though "[humans] make their own history . . . they do not make it just as they please; they do not make it under circumstances chosen by themselves, but under circumstances directly found, given and transmitted from the past."[44] The habitus concept provides the means to explain how systemic racism's indoctrination engenders resilient racial dispositions in a vast majority of genomicists that impact their conceptualization of 'race.'

Comprehending how racism plays out in the complex and difficult context of genomics is challenging but possible. An important starting point is to recognize that genomics, like all arenas of knowledge, is a conduit through which power is exercised.[45] As Michel Foucault observed, discourses produce knowledge and knowledge is always a weapon of power.[46] Over two centuries scientists created and then tried to confirm that 'race' is biologically real, thereby validating 'white' superiority, although, as I discuss later in the book, it is now known this 'scientific' exercise is pointless given that 'race' does not map well onto genes. Nonetheless, research stipulating disease susceptibility variation among socially defined racial groups results from distinct biological structure beyond the influence of social conditions remains prevalent in genomics.[47] Researchers who espouse such views include prominent geneticist Neil Risch and collaborators, and biologist Armand Leroi, who insist that self-ascribed 'race' and 'continental ancestry' are useful shorthand for speaking sensibly about genetic differences between social groups.[48] Exactly why some genomicists continue to look for bio-'race' given the voluminous evidence refuting its existence as we will see is a by-product of racial meanings that constrict their ability to comprehend genetic variation not as the normal character of human biology, but as a marker of biologically distinct differences. For this reason genomicists who generally assert that biological 'races' exist are susceptible to ignore "evidence that works against [their] belief" because of its importance to their racial conceptions of themselves and humanity.[49] To protect their core belief about the 'race' concept, they . . . rationalize, ignore and even deny anything that doesn't fit in with [it]."[50] Consequently, genomicists are generally inclined to accept and internalize characterizations of genetic variance as biologic racial distinctions. Although these racial 'truths' are not based in fact, they are thought of and acted upon as such.

Because systemic racism's ideological components distort genetic reality, I approach racial thought in genomics as facilitating action and discourse, whether intended or not, that solidifies 'race' as normative in genomics.[51] Consequently, I view racialized culture and ideology as mechanisms for imprinting 'race' onto the narrative of biological inheritance.

SOCIAL SCIENCE ON RACIAL IDEOLOGY IN GENOMICS

Although scholars are increasingly examining how scientific activity is interwoven with racial meanings, beliefs, and power, critical inquiry into the implicit racial practices, assumptions, and values informing genomics is only now starting to emerge.[52] Troy Duster's insightful analysis of the social and ethical implementations of modern molecular genetics provides a guidepost for identifying the interplay between genomics and racial culture, but it does not unravel how raced social relations create racialized genetic understandings.[53] While Jenny Reardon's analysis of the Human Genome Diversity Project (HGDP) expounds on Duster's analytical roadmap of the interaction between genomics and racial ideology, like Duster's study, it too does not adequately consider how 'race' as a social location and ideology becomes subsumed in researchers' ways of thinking about and conceptualizing human genome variation.[54] Her research does, however, demonstrate that genomic ideas and practices are not merely innocent tools for fighting racism. Genomics' embeddedness in consequential social and political environments means it is also plays a role in defining and shaping 'race' and racism. Genomicists' failure to recognize such social influences are not merely symptomatic of falling prey to ideological blinders. It also signals, as Reardon argues, that the present genomic moment is beset by a fundamental analytical weakness in how social scientists, historians, and scientists alike comprehend the role genomic ideas and practices play in creating and defining 'race.'[55]

This analytical deficiency exists in part because most genomicists erroneously presume genomics is not inextricably linked to racialized culture. It is simply impossible for either 'race' or genomics to construct one without being implicated in the construction of the other.[56] Lewontin maintains that such mutual construction occurs because:

> There [is] constant pressure from social and political practice and the coincidence of racial, cultural, and social class division reinforcing the social reality of race to maintain race as a human classification. If it were admitted that the category of race is purely a social construct, however, it would . . . weaken [its] legitimacy. Thus there [are] repeated attempts to reassert the objective biological reality of human racial categories despite the evidence to the contrary.[57]

Despite genomicists' contention, there is no neutral, apolitical ground on which they perform their research.[58] Reardon contends therefore it is vital for scholars to account for the racialized cultural process through which genomics creates molecular differences between people who are not really different from one another and labels people who are molecularly similar as different.

Duana Fullwiley examines how commonsense racial ways of knowing operating as normative genomic practices transmit notions that regenerate racial genetic homogeneity and intergroup distinction.[59] According to Susan Lindee and her collaborators, this racializing process is driven in part by the fact that "genetics in practice is plastic and contingent, embedded deeply in culture, time, and place."[60] Similar to the way the 'race' concept is used, the fluidity of genomic meanings gives it the capacity to conceptually alter itself to accommodate itself to the norms of existing racial order.[61] Within the racial status quo "genes [operate as] resources for many different groups, deployed to resolve longstanding disputes among [them], negotiate international trade, explain historical events inaccessible in any other way, and contest oppression and racism."[62] Catherine Bliss promotes an "anti-racist"[63] approach with her contention that it is possible for genomicists to engage the 'race' concept as a means to dispel erroneous assumptions and research specifying the existence of biological differences between socially defined racial groups.[64] This work is difficult because "anti-racist" genomicists, like everyone else in society, are subsumed in racist meanings that operate automatically—and often unconsciously—when constructing genome knowledge. Being in opposition to racism does not inoculate one from participating in it. There is no individual escape from whiteness because all socially defined racial groups are socially bound together by it. Systemic racism's omnipresence means, as philosopher Eric Juengst explains, "No matter how great the potential of . . . genomics to show our interconnections, if it begins by describing our differences it will inevitably produce scientific wedges to hammer into the cracks that already divide us."[65]

Michael Montoya's[66] research regarding the genetic etiology of diabetes among Mexicans differs from Bliss's[67] in that it argues that when genomicists use racial taxonomies to identify differences in susceptibility to diabetes, the social identities and life conditions of DNA donors are grafted into the biological explanations of human difference and disease causality. This collapsing of sociohistorical and biogenetic categories leads to biological variation being organized not around individuals with perceived differential genetic risks but rather stratified groups with perceived similar genetic differences.[68] Because 'race' is a construction of genomicists' ever-changing socially constructed racial imaginations and genotypes are stable features, Montoya maintains genes found within racially labeled groups cannot explain the variance in group susceptibility to diabetes.[69] This is so because genetic discoveries for diabetes among socially defined racial groups within high-tech genomic laboratories are based on finding similar genetic patterns within these groups to explain the onset of illness. Yes, genetic factors influence the incidence of diabetes but so too do a broad range of environmental and sociocultural factors (e.g., access to healthy food, health care, housing, etc.). Because 'race' as a sociocultural concept references social conditions, both historical and contemporary, of socially defined racial groups, Montoya contends that the quest to find genetic susceptibility diminishes the importance of political risk factors in creating diabetics.[70] Genomics' depoliticized approach inclines genomicists to perceive and present nongenetic racialized factors as intractable characteristics of marginalized socially defined racial groups that are unapproachable or too 'complex' for genomics to solve. Although it is recognized that the more credible explanation for diabetes susceptibility is found in the interplay between biology and social environment, the focus remains on locating only its genetic underpinnings through group comparison. In this comparative context, 'race' operates as the basis by which populations are chosen to study and a proxy for phenotype. Montoya's research moves us a step closer to clarifying how racial meanings mediate interpretations of genetic information. Alondra Nelson's research buttresses Montoya's findings with its demonstration of how users of genetic genealogy tests alter the uncertainty of genetic data to fit their socially contrived racial conception of reality.[71] In this way, her research also denotes the importance of social location in making sense of genetics.

Dorothy Roberts's[72] engagement with genome science's preoccupation with the biological basis of 'race' is similar to Reardon[73] in that she contends it is problematic to simply dismiss the racial science of the past as "pseudoscience." To deconstruct the myth of intrinsic racial differences and challenge the ideas that racial science no longer exists, Roberts distinguishes between accepting a false view of 'race' as an inherent biological category written in our genes and 'race' as a political system of governance invented to perpetuate systemic racism. According to Roberts:

> Race is a political grouping that uses various biological demarcations [to] distinguish who belongs to one or another. But those—skin color, hair color, the shape of the nose or the lips—are only part of what we use to determine what race someone is. We also historically have looked at their behavior, we've looked at who their friends are, where they live, to also help determine it. So there's a grab bag of biological, physical, social and cultural clues that we use every day to decide who belongs in what race.[74]

Similar to Roberts, I highlight the importance of understanding 'race' as a deeply embedded ideology many scientists automatically use in their research. It is virtually impossible for them to imagine another way of doing research because of their social conditioning—not only as scientists but as individuals who were racially socialized by their parents, peers, media, and education. As a consequence of this conditioning, Roberts argues that the field of genomics, like the twentieth century genetics, is predisposed to uncritically importing assumptions about the biological basis of 'race' into the discipline.[75] Though it is tempting for genomicists to contend racial science is unbiased in an attempt to valorize racial hierarchy, such hyperbole provides a way for them to present their studies of biological racial difference as color-blind—free of the racial bias characterizing the "pseudoscience" of an earlier period. But as Roberts correctly observes "what [we] call racial pseudoscience today was considered the vanguard of scientific progress at the time it was practiced."[76] Her point is that current genetic examinations of racial difference are not distinctly free from unsound empirical assumptions and implicit bias. As Roberts explains, "[t]he burning scientific questions of each period have been framed and answered in terms of race not because rational scientific inquiry compelled it, but because race was presumed to be an essential biological category."[77]

Roberts's research illuminates how knowledgeable and (assumedly) non-racist genetic scientists so easily adopt unsound assumptions about the scientific nature of 'race.' According to Roberts, the new racial genomics operates as a seemingly non-biased (color-blind) explanation for continuing racial inequality.[78] It hides the political nature of 'race' by recasting it as nothing more than a probabilistic grouping rather than a system of 'race'-based inequities.[79] That is to say, genomicists claiming to be philosophically and methodologically color-blind are averse to considering how racism induced institutional inequities impact human biology. Though some natural genetic divisions exist, human biology is influenced by 'race' because 'race' as a political division shapes institutions that treat people unequally.[80] Because many genomicists are opposed to openly accounting for how social definitions of 'race' inform genetics,

Roberts argues their reasoning and practices—which appear to be colorblind—merely reinvent 'race' as a biological reality. As we will see, by associating various genetic markers with 'biogeographical ancestry' (a fancy repackaging of the nineteenth-century racial categories) genomicists refashion 'race' as scientific truth at the genetic level. In this way, Roberts maintains "biological distinctions, seemingly validated by genomic science and technology, appear to explain why stark racial disparities persist despite the abolition of official discrimination on the basis of race and despite most white Americans' belief that racism has ceased to exist."[81]

Following Roberts, Reardon, and Duster, I investigate how racial ideology subsumed within genomics creates a molecular rationale for the biological basis of 'races' that appears non-racist. Detailing the social processes through which racialization occurs in genomics requires an exploration of genomicists' actions and thinking in relation to others, since it is through their social relations they come to practice science as they do. Genomicists' enculturation in racialized structures of the family, state, and production strongly influence how and what they think or do not think, and culturally bend genomic research to conform to a white supremacist version of reality. Examining the links between racial ideology and genomics means going well beyond investigating an individual's racial biases to consider how 'race' as a cultural constituent conditions genomicists to perceive, think, and act as if whiteness is 'race' neutral/ordinary. An accurate understanding of how 'whiteness' is taken for granted requires accounting for how genetic and social history impacts ongoing structural and institutional racial realities relevant for comprehending contemporary genomic developments. To this end, I document how these structural and institutional racial processes constitute genomics' description of human genome diversity in ways that are congruent with the material and mental reality of systemic racial oppression.

EXPLANATORY FOUNDATION

Social Constructionism

Scholars who examine how genome science, racialized culture, and ideology interrelate generally use the theory of social constructionism. Social constructionism originates in the work of George Herbert Mead (1934) and the writings of scholars like Berger and Luckmann (1967) as an attempt to grasp how meanings about reality are created, negotiated, sustained, and modified through people's social interaction with one another. Since reality is open to interpretation social constructionism contends that people establish implicit social agreements about how to construe facts and events. This agreed upon basis for making sense of physical or objective reality is known as social reality. This consensus-base view of knowledge about reality means that it is not the state of things as they actually exist in the world but rather products of our socially shared understandings of how they may appear or might be imagined. The point is, reality for social constructionism is a creation of shared experiences in the world about how reality should be understood thus people's experiences with the world do not accurately describe reality.

Since the material world can only be known in relation to people's experience within it, we constantly refine our ideas about reality as we acquire more experience with the world. In this regard people's simultaneous existence in both objective and subjective realities means that our interpretations of objective reality—facts and truths—are creations of social agreements established in specific cultural, economic, political, and historical moments and settings. Since our shared beliefs and descriptions about reality make it comprehensible, knowledge production is inherently a social creation, not a discovery of the mind.[82] Because scientific inquiry is a contingent social activity it is important for scientists to acknowledge science's social creation to account for how other aspects of their social lives affect its production.[83] Although constructionism asserts that specific, historically contingent community interests and goals inform science, it views reality as independent of the meaning scientists give it.[84] Because we cannot examine reality—objectively and scientifically—without interpreting it through our own culturally inherited, ubiquitous frames of reference, the theory of social constructionism holds knowledge creation is a social endeavor influenced by specific racial histories.[85]

Genomicists internalize the culture in which they are situated as their own through primary and secondary socialization. Primary socialization provides individuals with identities and social scripts for roles assumed in society. Through social interaction with significant others within specific racialized sociopolitical contexts, genomicists acquire vast pools of transmitted experiences from others which they use to construct novel racial ways for understanding genome reality.[86] In this respect, genome knowledge construction and cultural processes are connected and interdependent. The unification of nature and culture in knowledge creation means genomicists' interpretation of the genome is a product of their physical interaction with nature and other humans. Because socialization is facilitated through interaction with others within specific sociohistorical and political contexts using language—spoken and unspoken—generated out of these environments, constructionism contends language is not merely a means for transmitting thoughts and feelings; it provides genomicists the means to structure the way the world is experienced.[87] It is through language that genomicists maintain, alter, and reconstitute their definitions of genetics so that their conceptions of genetic processes are collectively shared in assumed ways that do not require their re-creation anew each time they are used to comprehend the genome. Once these collectively held conceptions are adopted it is difficult to change them because people internalize them as their own.

Genomic debates regarding the usefulness of the 'race' concept for inferring human genetic variation are entrenched in racial ideology.[88] These debates reveal how genomics sustain the 'race' concept by generating multiple meanings to promote its viability as a scientific concept.[89] The persistence of the 'race' concept in genomics is particularly palpable among researchers who simultaneously acknowledge that it is an imperfect sociohistorical construct that is transmuted into biology while using it as a proxy to infer genetic variation. In the book I contend that their use of 'race' within raced cultures unavoidably preserves the fallacious assumption that genetic diseases are racially distinct. Social constructionism is applied to show that genomicists' failure to consider how their means for conceptualizing genetic diversity results from their internalization of racial ideology inclines many of them to view and reproduce 'race' as a valid proxy for inferring biodiversity among social groups. This seems to be what occurred in the study of Tang and her colleagues, which extolled the usefulness of 'race' in clinical studies.[90] Their categorization of 3,636 people self-identifying as members of one of the four major racial groups ('white' Euro, 'black,' Asian, and Latino) in the United States suggests genetic variation correspond to these racial categories. Rather than partition genetic variation—most human variation occurs within socially defined racial groups rather than between them—Tang and her colleagues *looked for* allele[91] clustering differences between socially defined racial groups to confirm 'race' as biologically real. Since the bulk of human variation is continuously distributed, sociologist Reanne Frank argues that any categorization schema that attempts to racially categorize variation will necessarily create artificial truncations.[92] This is the primary reason why I argue that attempts to allocate individuals into ancestry groupings based on genetic information yields varying results that are highly dependent on methodological design.

Although it is possible to find genetic variation roughly corresponding with 'race' based on different allele frequencies, this is not the case for all geographically distinct populations.[93] Evolutionary biologist Joseph Graves argues that Tang and her colleagues found racial clustering because they assumed that self-identified 'race' maps to discrete racial categories when in fact the allele variations were merely mirroring the genetic distance between populations. Because people mate primarily within groups they share allele frequencies that cluster around kin groups and lineages, or by national, cultural or linguistic boundaries—that correlates well with geographical groupings. However, biological variation of any single human genetic trait is described best as a gradual change in trait frequencies across the distributional geographical range of humans. Tang and her collaborators' decision to sample distant continental racial groups in the United States ensured acute genetic geographical discontinuities fell along perceived racial divides. This occurred because genetic data clustering is highly dependent on the genomicists' initial racially distorted hypotheses and methodologies. Because genetic data are chosen and analyzed by genomicists immersed in racialized cultures that provide them the means for making sense of human genetic variation they, as I discuss later in the book, are apt to design studies and interpret data to suggest a relatively small fraction of genetic variation between socially defined racial groups connotes biological 'race.' Although socially defined racial groups have different allele frequencies, this does not mean that racial classification is a natural taxonomy within the human species, but rather genomicists' failure to problematize how the racial ideology subsumed their genomic thinking. Social constructionist theory's premises about social conditioning and racial situatedness are used to delineate how racial ideology informs genomicists' conceptualization and understanding of the human genome variation.

Standpoint Theory

Feminist standpoint theory, whose theoretical foundation lies with Karl Marx's thinking about the social construction of consciousness, expounds on social constructionism assertions regarding cultural situatedness by adding individuals' social position in society also informs what they do and how they think about reality.[94] Adherents of standpoint theory contend that the power relations that structure society help shape genomicists' perceptions of the genome. As such, these matrixes of domination and exploitation are central to understanding the human genome; genomicists situated in different social structural locations (i.e., racial, ethnic, class, gender, etc.) yield a variety of views about genetic reality. Although standpoint theory recognizes that all genome knowledge is socially situated, its exponents assert that not all social positions are equally good locations from which to observe and explain human genome variation. Thus all genomicists within the context of systemic racism are prone to cultivate ways of knowing that perpetuate rather than allay white supremacy. In other words, genomicists' knowledge claims are creations of specific sociohistorical conditions and contexts; their learned ways of knowing reality are creations of the sociopolitical environment in which they are situated. Genome knowledge is not merely created in idealized logical space; it is conceived within a social and power nexus. As a result, it matters who determines which genomic concepts are intelligible, which genomic claims are heard and understood by whom, which genome features are considered important, and which genomic explanations and reasoning are taken as credible. Although genomicists try to craft genomic explanations devoid of social influences, genome science's proclamation of objectivity is but a theoretical ideal and "in practice objectivity is a standpoint—[elite], white, and male."[95] My own argument draws on standpoint theory to investigate how human genome science, as dominantly practiced, offers partial views of genome reality that construct 'truths' serving dominant group interests.[96] Genomic members of the social and political dominant 'white' clique are more inclined to exclude questions about genetic diversity challenging their group's privileged position than genomicists from subordinate socially defined racial groups. This predilection is even evident among 'insurgent' 'white' feminist analyses of science whose critiques of science tend to routinely disregard science's role in making 'race.'[97] To overcome this tendency, standpoint theory urges genomicists to consider how their life experiences manifest in their thinking and research. This requires genomicists to place their "class, race, culture and gender assumptions, beliefs and behavior . . . within the frame of [biological reality] s/he attempts to paint. . . . Thus the researcher appears . . . not as an invisible, anonymous voice of authority, but as a real, historical individual in concrete, specific desires and interests."[98]

Scholars critical of standpoint theory assert its emphasis on the universality of group structural experiences comes at the expense of not accounting for the differences found between subordinate groups' understanding.[99] Sociologist Patricia Hill Collins insists, for example, that standpoint theory poorly addresses how marginalized groups' sites of counter-discourse are themselves contested terrain where power differences between and among these groups influence how reality is perceived and interpreted.[100] Despite the multiplicity of standpoints and power relations, standpoint theorists hold firm to the view that knowledge production is subjective in the sense it is rooted in particular subject positions (bourgeois or proletarian, 'white,' male, etc.), thus these subjective positions are creations of systems of oppression.[101] Standpoint theory's emphasis on the universality of the oppressed's standpoint, however, prevent it from considering how subordinate socially defined racial groups' internalization of prevailing racial values and ways of knowing mold their thinking about reality. For dominant racist ideology and discourses overlay standpoint epistemologies. Despite this shortcoming, standpoint theory's approach to knowledge production as an ongoing creation of historical oppressive relations and positions is helpful in clarifying how genomicists' racial experiences, knowledge, and identities are insinuated in genomic knowledge. Such analytical illumination is crucial to explaining how and why many genomicists, no matter their group affiliation, are often oblivious to how white supremacy shapes their language and conceptions of human biological differences.

Co-production Theory

Along with social constructionism and standpoint theory, co-production theory is also used to analyze genomics as neither a simple reflection of nature nor an epiphenomenon of social and political interests. This theoretical approach provides my analysis with a more dynamic interactive systemic frame for deducing how genomics shapes and is shaped by racialized culture, with genomic rationalization interconnecting with racial reasoning.[102] In practice this means that genome knowledge and 'race' are produced by people in scientific institutions and cultures containing built-in biases that naturalize racialized meaning and understandings. Understanding the collective construction of genome knowledge requires inquiry into how contingent social conditions affect the production and evaluation of genomic accounts.[103] Myopic narratives presenting science as a purely social product with any relation to reality, or explaining racism as simply the behavior and attitude of individuals, inhibits our understanding of how genome variation discourse operates simultaneously as a scientific and racial discourse. Co-production theory assumes that genomic knowledge is not constituted any differently from reasoning in society. As such, the two are inseparable because they are cut from the same cultural cloth.[104] As Lewontin's *Biology as Ideology* reveals, social norms and expectations embody researchers' reasoning regarding genetic properties.[105] Lewontin argues that this is evident when genomicists devoted to the bourgeois idea of individualism use it to frame genetic disease susceptibility as an outcome only of similar molecular essence among social groups rather than groups' varied exposure to socioenvironmental forces that alter the genetic structure in ways that increase or decrease a group's vulnerability to disease.[106] Genome science does not occur in a social vacuum. Co-productionism makes it plain that genomicists' interpretations of biological reality are (re-)constructing the social as they (re-)construct knowledge about genome variation.

A common criticism of co-productionism, much like its social constructive counterparts (social constructionism and standpoint theory), is the idea that it asserts that scientific knowledge is entirely socially contingent. This is not its claim. Rather, co-production theorizes that knowledge is not subjective in that individuals and groups cannot impose their own point of view without having to relate it to the perspectives of others, particularly those of the dominant group. What co-productionism advocates is not the relativity of truth but rather the truth of social relations in knowledge creation. Since knowledge of objective reality is produced within specific social and historical contexts, co-productionism assumes scientific reference frames about reality shift in point of view, making it possible to observe how particular sociopolitical contexts and processes shape explanations and representations over time.[107] By distinguishing relatively, not absolutely, between objects of study 'as they are' and the representation individuals make of them, co-production theory helps my analysis explain how genomic interpretations and understandings are constructed in relationships with other social phenomena and power. The language genomicists use does not merely reflect the way biological reality actually is, rather it is a description of how biology is made by scientists, who in turn, are social products of the descriptive categories that they internalize and use.[108] I use co-production theory's premise that understandings of biological reality are sociohistorically constituted thus predicated on established cultural and power considerations to interpret the hierarchical social processes through which genome science acquires racial form and meaning.

Systemic Racism Theory

Systemic racism theory emerged from the thinking of activists and intellectuals such as Frederick Douglass, Ida B. Wells-Barnett, W. E. B. Du Bois, Oliver Cox, Anna Julia Cooper, Kwame Ture, and Frantz Fanon to contend that social systems in racialized societies imbed racism at their "very core."[109] Unlike the dominant societal view that racism results only from individual racial prejudice, the systemic view of racism sees it as operating imperceptibly as a large-scale inescapable hierarchical system of domination generating and sustaining itself as the material, social, and ideological reality within societal institutions.[110] Systemic racism involves various institutional decisions and policies working in the interest of 'whites' at the expense of other socially defined racial groups. My analysis uses key aspects of systemic racism theory to assess the impact of racism on the thinking and actions of genomic researchers. Absent a systemic racism analysis which goes beyond individual racism, genomics is unlikely to seriously address the racist realities it helps to perpetuate.

Sociologist Joe Feagin, a contemporary proponent of systemic racism theory, specifies the concept as delineating various dimensions of racism: the (1) dominant racial hierarchy, (2) comprehensive white racial framing, (3) individual and collective discrimination, (4) social reproduction of racial and material advantage, and (5) racist institutions integral to white domination.[111] The conceptual language of the second and fifth dimensions of racism are used in the book to examine how most genomicists articulate 'white' racial framed discourses like the notion that racism is an individual problem of a few bigots to divert focus away from their role in propagating systemic racism. As I detail in the book, such redirecting is common in genomics according to the few racially oppressed genomicist interviewees who maintained it was driven largely by ignorance and sometimes willful ignorance of racial undercurrents in genomics.

Given racialized culture's role in informing the way genomicists create, view, speak, and write about the human genome, systemic racism theory helps in explaining how 'whites' in particular play a direct and ongoing role in producing and promoting white supremacy. 'White' genomicists' immersion in a racialized culture ensures they and other marginalized group members use what Feagin terms "white-centered frames" that keep them from seeing and/or acknowledging their role in creating and recreating systemic racism.[112] The 'white' racial frame, as a critical aspect of systemic racism, functions as a generic meaning system, transmitted and held by most 'whites' and, at least in part, accepted by racially oppressed people. Among 'whites' in particular, this framing is deep-seated, broad, and encompasses many pieces of racialized knowledge and understandings that shape human action and behavior in a myriad of ways, often automatic or unconscious. As a deeply engrained idea, the 'white' racial frame is more than cognitive; it contains racial images, interpretations, emotions, and action inclinations that are closely linked to genomicists' racial cognitions and understandings. In Feagin's words, 'white' racial framing informs the "language chosen to describe a society demographically or sociologically often reveals white perspective."[113] This means that prevailing ways of talking and thinking about racial concerns are reflections of the hegemonic 'white' racial frame. Systems of 'white' racial frames are not merely bodies of ideas but lived material practices—ways of thinking taking practical form—reproduced through the practices and in social and knowledge activities institutions like science. Although genomics presents itself as a site of 'objective' reasoning, it and its practitioners are in reality products of racism. For example, genomics' focus on finding racial genetics is in many respects related to its embeddedness in systemic racism, which encourages genomicists to categorize individual genetic variance as racial group differences. This propensity for identifying difference as indicative of racial biological distinction produces racial 'truths' by which people organize their lives.

Systemic racism theory is a useful interpretation for my study because it recognizes that racial ideology and oppression change and adapt to external shifts and resistance pressures without significantly altering the essential structures and foundations of racism. With this theoretical premise we can account for, for example, why eugenics and its racial ideology did not fade away when its popularity waned after Nazi Germany's eugenics was discredited in the 1940s. Instead, as I explain in chapter 1, it tethered itself to the rising field of genetics. My argument draws upon systemic racism theory to conceptualize how racism structures genomics and researchers' thinking and practices to prohibit their acknowledging the existence and operation of systemic racism within genomics. In short, I attempt to grapple with and clarify the intransigent character of racism and racial ideology as multivalent systems of discursive and material practices structuring genomics.

STUDY DESIGN

Systemic racism theory, constructionism, standpoint theory, and co-productionism are used in the book to explain how racial ideology which is naturalized and made invisible in genomicists' investigative routines shapes genomic research. I contend that it is impossible to comprehend the persistence of racial thinking in genomics without taking into account how socially constructed racial definitions and understandings mediate genomicists' understanding of objective genetic variance. To determine if my contention is plausible, the book draws on interviews I conducted, archival interview data, and existing qualitative research. The archival data consist of thirty-five in-depth interviews with genomic researchers conducted from 1996–2015 while the primary data involved forty-three in-depth interviews with genomicists from 2009–2012 from various subfields (ranging from molecular biology to neurogenetics). The interview data included a racial and gender diverse set of sixty-three self-identified males and fifteen women, and sixty-eight self-identified 'white' Euro-Americans, five Asian Americans, four African Americans, and one Latino American. Interviews conducted for the study were gathered chiefly through telephone, video chats, and face-to-face conversations. Primary interviews lasted from forty-five minutes to an hour and consisted of open-ended questions regarding their thoughts about their conceptualization of 'race's' cultural and social influences on genomics (see Appendix A for interview questions).

To substantiate how racial meanings are implicated in genomics, the book used ethnographic content analysis to convert interview data into descriptions of emergent patterns, emphases, and themes that access underlying meanings and processes at the intersection of racialized culture and genomics.[114] Ethnographic content analysis focuses on the dynamic interplay between social contexts, genomic knowledge construction, and discourse, rather than mere quantity and/or numerical relationship between two variables. This provided me with the means to track racialized genomic discourse, themes, and frames over time, across different genomic issues, and in different social settings.[115] I used this methodological process in my analysis of the interviews to reflexively move between concept development, sampling, data collection, coding, analysis, and interpretation to reveal the themes and theories emerging from genomicists' discussion. As an interpretive tool for evaluating meaning in particular, sociohistorical contexts, ethnographic analysis required that I be reflective about my biases as I interpreted the data. To this end, while gathering and analyzing the interviews, I reflected on how my social position and identity as a middle-aged African American male social scientist affected interviewees' answers and my interpretation of the data. In this way, I remained thoughtful and forthright about tensions surfacing while I collected and analyzed interview data.

Discourse analysis was used with ethnographic content analysis to probe beneath the surface meanings of interviewees' spoken words and interpretive frames to uncover the underlying relations of racial power infused in their 'talk' about genomics. Discourse analysis gives the study the capacity to evaluate the actual ways genomicists produce practices and discourses within racially constraining sociocultural environments. More precisely, discourse analysis allowed me to consider how preexisting racial discursive frames and cues, conscious or unconscious, can induce genomicists to create genomic explanations that are congruent with commonsense racial understandings. To discursively frame is to select some aspect of perceived genetics to be most salient, and useful for conveying genomic understanding promoting a particular definition, causal interpretation, moral evaluation, and/or treatment recommendation of the genome described.[116]

Both discourse and ethnographic content analysis provided me with the methodological means to approach the interviewees not as passive recipients of fixed racial meanings, but as individuals ontologically and epistemologically equipped to produce and possess fluctuating meanings about human genome variance that are more or less congruent with societal racial conceptions of reality. In short, I used discourse and ethnographic content analysis in the book to make transparent how interviewees' racial ways of knowing blends into their conception of human genome variation. Knowing exactly how this social process works is crucial because it is through the flow of racialized human interaction that genomics as a field emerges.[117]

OUTLINE OF CHAPTERS

Although the human genome exists apart from society, knowledge about it is produced through socially created language and interactions. As such, genomicists' thinking is informed by their inability to escape the wake of the 'race' concept. This book is concerned with how racism makes genomics and how genomics makes racism and 'race,' and the consequences of these constructions. Specifically, I explore how racial ideology works in genomics. The simple assumption that frames the book is that 'race' as an ideology justifying a system of oppression is persistently recreated as a practical and familiar way to understand biological reality. The book reveals that genomicists' preoccupation with 'race'—regardless of good or ill intent—contributes to its perception as a category of differences that is scientifically rigorous.[118] In the next two chapters (1 and 2) I examine in detail the historical links between racial ideology, genetics, and genomics so that we may grasp how the researchers' signification of 'race' moved from phenotype to genotype. Chapter 1 presents relevant historical background for understanding the depth of racialized thinking's embeddedness in genomics, while chapter 2 explores the socialized process through which genomicists internalize enduring racial norms and values that inform their genomic thought and practices. Chapter 3 examines how racialized culture mediates genomicists' understandings and representations of objective genetic variation. In chapter 4 I discuss how racialized thought is made invisible through genomicists' objectivity claims and heuristic practices. I investigate how genomicists' 'bad science' discourse obscures racial thinking in their methodologies and efforts to create personalized genomic medicine in chapter 5. I then conclude in chapter 6 with a brief summary of my findings in which I urge genomicists to account for and lessen racial ideology's influence in genomics by centering it their research.

Collectively, these chapters demonstrate clearly the complexity of racism, racialized culture, and racial ideology and the wide scope of their impact on genomics, particularly genome variation research. At the same time, however, they reveal the imprecision and conjectural quality of genomicists' understanding of the impact of racism and racial ideology in human genome knowledge production.

NOTES

1. Mannheim, *Ideology and Utopia*; Žižek, *How to Read Lacan*, 16.
2. 'White' is a philosophically problematic term, particularly in regard to multiraciality and intersectionality. But the term is sociologically important because of the way it is routinely utilized, most especially by self-defined 'whites' to reinforce societal racial order.
3. Kuhn, *The Structure of Scientific Revolution*.
4. 'Race' is encased in quotations because it is a problematic cultural invention of arbitrary meanings applied to what appears as natural divisions within the human species. Race is an ideology that structures and controls discourse in support of white supremacy.
5. Lee, "Biobanks of a 'Racial Kind': Mining for Differences in the New Genetics," 443–460.
6. Wilson, *Racism: From Slavery to Advanced Capitalism*, 25.
7. Wilson, *Racism: From Slavery to Advanced Capitalism*, 25; Joe R. Feagin, *The White Racial Frame: Centuries of Racial Framing and Counter-Framing*.
8. Wilson, *Racism: From Slavery to Advanced Capitalism*, 25.
9. Bonilla-Silva, *Racism Without Racists: Color-blind Racism and the Persistence of Racial Inequality in the United States*.
10. The relationship between anti-racism and the concept of whiteness is complex and contradictory. Anti-racists place the critique of 'white' Euro-American racial attitudes at the center of their effort and in doing so ignore the historical and structural workings of racism. But anti-racism does urge us to question and deconstruct all racial categories including 'white.' Whilst the latter ambition represents an anti-essentialist rejection of the racialization process, the former reflects and reproduces monolithic, reified constructions of whiteness (see Bonnett, "Anti-Racism and the Critique of 'White' Identities."
11. Lipsitz, *The Possessive Investment in Whiteness: How White People Profit from Identity Politics*.

12. Roberts, *Fatal Invention: How Science, Politics, and Big Business Re-create Race in the Twenty-first Century*.
13. Roberts, *Fatal Invention: How Science, Politics, and Big Business Re-create Race in the Twenty-first Century*.
14. It is important to note that genomics is mainly a method for evaluating the function and structure of an organism genome that emerged from the discipline of population genetics.
15. This division of human physical variation into 'races' was undertaken mainly to justify the kidnapping and enslavement of Africans and genocide of indigenous people for their land.
16. Desmond and Moore, *Darwin's Sacred Cause: How a Hatred of Slavery Shaped Darwin's Views on Human Evolution*.
17. Desmond and Moore, *Darwin's Sacred Cause: How a Hatred of Slavery Shaped Darwin's Views on Human Evolution*.
18. Graves, *The Emperor's New Clothes: Biological Theories of Race at the Millennium*.
19. Graves, *The Emperor's New Clothes: Biological Theories of Race at the Millennium*.
20. Graves, "Evolutionary versus Racial Medicine," 142.
21. Graves, "Why the Nonexistence of Biological Races Does Not Mean the Nonexistence of Racism."
22. Graves, "Why the Nonexistence of Biological Races Does Not Mean the Nonexistence of Racism."
23. Graves, "Why the Nonexistence of Biological Races Does Not Mean the Nonexistence of Racism."
24. Graves, "Why the Nonexistence of Biological Races Does Not Mean the Nonexistence of Racism."
25. "Population" is encased in quotes as discussed later in the book because it operates in genomics as synonym for 'race.' See: Graves, *The Race Myth: Why We Pretend Race Exists in America*, 111–112.
26. Dobzhansky, *Genetics of the Evolutionary Process*, 355.
27. Graves, "Why the Nonexistence of Biological Races Does Not Mean the Nonexistence of Racism."
28. Mannheim, *Ideology and Utopia*, 29.
29. Mannheim, *Ideology and Utopia*.
30. Berger and Luckmann, *The Social Construction of Reality*.
31. Louis Althusser, "Ideology and Ideological State Apparatuses (Notes towards an Investigation)," 85.
32. Mills, *The Racial Contract*.
33. Reed, Jr., "Marx, Race, and Neoliberalism.'
34. Lewontin, *Biology as Ideology: The Doctrine of DNA*.
35. Templeton. "Human Races: A Genetic and Evolutionary Perspective."
36. Mills, *The Racial Contract*.
37. Feagin, *Racist America: Roots, Current Realities and Future Reparations*, x.
38. Mills, *The Racial Contract*, 126–127.
39. Feagin, *Systemic Racism: A Theory of Oppression*.
40. Gramsci, *Selections from the Prison Notebooks of Antonio Gramsci*, 377.
41. Althusser, *For Marx*, 233.
42. Hall, "Signification, Representation, Ideology: Althusser and the Post-Structuralist Debates."
43. Bourdieu, *Outline of Theory and Practice*.
44. Tucker, *The Marx-Engels Reader*, 595.
45. Spears, "Negotiating Racism in the Academy," 146–154.
46. Foucault, *Power/Knowledge*.
47. Rose, "Introduction to the Discussion of Race and Ethnicity in Nature Genetics."
48. Risch et al., "Categorization of Humans in Biomedical Research: Genes, Race, and Disease"; Armand Leroi, "A Family Tree in Every Gene," A21.
49. Fanon, *Black Skin, White Mask*.

50. Fanon, *Black Skin, White Mask*.
51. Spears. "Negotiating Racism in the Academy."
52. Benjamin, *People's Science: Bodies and Rights on the Stem Cell Frontier*; Bliss, *Race Decoded: The Genomic Fight for Social Justice*; Roberts, *Fatal Invention: How Science, Politics, and Big Business Re-create Race in the Twenty-first Century*; Fullwiley, "The Biologistical Construction of Race," *Social Studies of Science*; Fullwiley, "The Molecularization of Race: Institutionalizing Human Difference in Pharmacogenetics Practices"; Reardon, *Race to the Finish: Identity and Governance in an Age of Genomics*; Santos and Maio, "Race, Genomics, Identities and Politics in Brazil"; Duster, *Backdoor to Eugenics*.
53. Duster, *Backdoor to Eugenics*.
54. Reardon, *Race to the Finish: Identity and Governance in an Age of Genomics*.
55. Reardon, "Decoding Race and Human Difference in a Genomic Age."
56. Reardon, *Race to the Finish: Identity and Governance in an Age of Genomics*, 162.
57. Lewontin,"Confusions about Human Races."
58. Reardon, *Race to the Finish: Identity and Governance in an Age of Genomics*, 163.
59. Fullwiley, "The Biologistical Construction of Race"; Fullwiley, "The Molecularization of Race: Institutionalizing Human Difference in Pharmacogenetics Practices."
60. Lindee et al., "Introduction: Anthropology in the Age of Genetics—Practice, Discourse, and Critique."
61. Rabinow, "Artificiality and Enlightenment: From Sociobiology to Biosociality."
62. Lindee et al., "Introduction: Anthropology in the Age of Genetics—Practice, Discourse, and Critique."
63. There are many different problems with the term 'anti-racist' but one of the most prominent issues is its focus on defining 'white' people in exceptional terms. They are 'white,' yet 'anti-racist.' As if participation in struggle for racial justice cancels out whiteness, privilege, and position within white supremacist hierarchy (see Park, "Challenging Racism and the Problem with White 'Allies': A Conversation with David Leonard," for a more involved discussion of the anti-racism concept).
64. Bliss, *Race Decoded: The Genomic Fight for Social Justice*.
65. Annas, "Genism, Racism, and the Prospect of Genetic Genocide."
66. Montoya, "Bioethnic Conscription: Genes, Race, and Mexicana/o Ethnicity in Diabetes Research."
67. Bliss, *Race Decoded: The Genomic Fight for Social Justice*.
68. Fujimura et al., "Introduction: Race, Genetics, and Disease: Questions of Evidence, Matters of Consequence."
69. Montoya, "Bioethnic Conscription: Genes, Race, and Mexicana/o Ethnicity in Diabetes Research."
70. Montoya, "Bioethnic Conscription: Genes, Race, and Mexicana/o Ethnicity in Diabetes Research"; Benjamin, *People's Science: Bodies and Rights on the Stem Cell Frontier*.
71. Nelson, "Bio Science Genetic Genealogy Testing and the Pursuit of African Ancestry."
72. Roberts, *Fatal Invention: How Science, Politics, and Big Business Re-create Race in the Twenty-first Century*.
73. Reardon, *Race to the Finish: Identity and Governance in an Age of Genomics*.
74. Desmond-Harris, "The Biggest Lie About Race? That Its Real."
75. Roberts, *Fatal Invention: How Science, Politics, and Big Business Re-create Race in the Twenty-first Century*.
76. Roberts, *Fatal Invention: How Science, Politics, and Big Business Re-create Race in the Twenty-first Century*, 28.
77. Roberts, *Fatal Invention: How Science, Politics, and Big Business Re-create Race in the Twenty-first Century*, 28.
78. Roberts, *Fatal Invention: How Science, Politics, and Big Business Re-create Race in the Twenty-first Century*.
79. Roberts, *Fatal Invention: How Science, Politics, and Big Business Re-create Race in the Twenty-first Century*, 291.

80. Roberts, *Fatal Invention: How Science, Politics, and Big Business Re-create Race in the Twenty-first Century.*
81. Roberts, *Fatal Invention: How Science, Politics, and Big Business Re-create Race in the Twenty-first Century*, 297.
82. Schwandt, "Three Epistemological Stances for Qualitative Inquiry: Interpretativism, Hermeneutics and Social Constructionism."
83. Maria Baghramian, "Constructed Worlds, Contested Truths."
84. Latour, *Pandora's Hope: Essays on the Reality of Science Studies.*
85. Burr, *Social Constructionism.*
86. Berger and Luckmann, *The Social Construction of Reality.*
87. Burr, *Introduction to Social Constructionism.*
88. Risch et al., "Categorization of Humans in Biomedical Research: Genes, Race, and Disease"; Royal and Dunston, "Changing the Paradigm from 'Race' to Human Genome Variation"; Long and Kittles, "Human Genetic Diversity and the Nonexistence of Biological Races."
89. Stevens, "Racial Meanings and Scientific Methods: Changing Policies for NIH-Sponsored Publications Reporting Human Variation."
90. Hua Tang et al., "Genetic Structure, Self-Identified Race/Ethnicity, and Confounding in Case-Control Association Studies."
91. Alleles are corresponding pairs of genes located at specific positions in the chromosomes. Together, they determine the genotype of their host organism.
92. Frank, "What to Make of it? The (Re)emergence of a Biological Conceptualization of Race in Health Disparities Research."
93. Graves, *The Emperor's New Clothes.*
94. Marx, *The Economic and Philosophic Manuscripts of 1844.*
95. Walker et al., "Inverting the Inverted Pyramid: A Conversation about the Use of Feminists Theories to Teach Journalism."
96. Harding, *Whose Science? Whose Knowledge? Thinking from Women's Lives.*
97. Collins, "Moving Beyond Gender: Intersectionality and Scientific Knowledge."
98. Harding, "Introduction: Is There a Feminist Method," 9.
99. O'Leary, "Counteridentification of Counterhegemony? Transforming Feminist Standpoint Theory."
100. White, "Africa on My Mind: Gender, Counter Discourse and African American Nationalism."
101. O'Leary, "Counteridentification of Counterhegemony? Transforming Feminist Standpoint Theory."
102. Lock, "The Tempering of Medical Anthropology: Troubling Natural Categories."
103. Shapin, "History of Science and Its Sociological Reconstructions."
104. Jasanoff, "A New Climate for Society."
105. Lewontin, *Biology as Ideology: The Doctrine of DNA.*
106. Lewontin, *Biology as Ideology: The Doctrine of DNA.*
107. Bova, "On Relativism, Pragmatism and Critical Theory."
108. Rorty, *Philosophy and the Mirror of Nature.*
109. Feagin, *Systemic Racism: A Theory of Oppression.*
110. Feagin, *Systemic Racism: A Theory of Oppression.*
111. Feagin, *Systemic Racism: A Theory of Oppression.*
112. Feagin, *The White Racial Frame: Centuries of Racial Framing and Counter-Framing*, 7.
113. Feagin, *The White Racial Frame: Centuries of Racial Framing and Counter-Framing*, 7.
114. Altheide, "Ethnographic Content Analysis," 65–77.
115. Altheide, "Ethnographic Content Analysis," 65–77.
116. Entman,. "Framing: Toward Clarification of a Fractured Paradigm," 51–58.
117. Geertz, *The Interpretation of Cultures.*
118. Holloway, *Legal Fictions: Constituting Race, Composing Literature.*

ONE

Genomics' 'Race' Legacy

Despite the scientific consensus that there are no human 'races,' socially created racial categories in genomics are often treated as if they are biologically real. This occurs in part because researchers exist within a racist context that instantiates racial ideology in their consciousness through endemic racialized practices within the institutional structures of society and genomics.[1] Genomics and genomicists are products of historic racial formations that define and give the human genome meaning. These meanings in turn are shaped by racialized genomic practices. Moreover, in genomics' racialized social setting, racial comparative genetic analyses assist in sustaining genomicists' racial dispositions. To illustrate how this comparative process works, consider that there were no blacks or 'whites' in history until Africans from different ethnic groups from across the continent were enslaved by Europeans, their enslavement fashioned the 'white' identity. For 'white' identity is dependent on there being an assumed inferior black 'other.' Given the centrality of this comparative racial typology for understanding the human genome, I argue in this chapter that it is crucial for genomicists not merely to think but to act their way into less racialized ways of thinking about genomics. However, before this occurs, genomicists and the public need to come to terms with the fact that the knowledge genomics produces is freighted to racial history. To this end, let us turn our attention to explaining how the early twentieth century racial ideology of eugenics coupled with genetics to extend the fictive notion of biological 'races' to contemporary genomics.

THE EUGENICS CONNECTION

The new science of genetics arose after Gregor Johann Mendel's work on heredity appeared in the *Journal of the Brünn Natural History Society* in

1866, in a paper detailing variation, heredity, and evolution in plants. In this paper Mendel demonstrated how pea plant characteristics are preserved as they pass through successive generations. Mendel's paper showed reproduction in organisms is a conduit for introducing genetic variation. In sweet peas as in humans, each plant is either male or female and possesses discrete units of inheritance that Mendel called "factors" (later labeled chromosomes).[2] Each offspring inherited two copies of their parents' chromosomes, one from each parent. When Mendel crossed tall pea plants with short pea plants the resulting offspring were not medium in height but rather almost uniformly tall. Crossing chromosomes produced a mathematically predictable pattern of inheritance that fit perfectly with Charles Darwin's theory of natural evolution, which up until Mendel's publication had no reasonable explanation for how biological adaptations to environmental conditions passed from generation to generation. Though Mendel published his research a mere seven years after Darwin's *The Origin of Species* appeared, Darwin was unaware of his work. This was not the case for Mendel regarding Darwin's ideas. He was very familiar with Darwin's work and aware that random variation inheritance is powerful raw material for demonstrating how natural selection works.

Mendel's research languished in obscurity until the end of the nineteenth and early twentieth century when 'scientific racism'[3] proponents rediscovered it. The reemergence of Mendel's work led to the creation of the new, exciting branch of biology called *genetics,* which scientific racists used to argue that the intermarriage of 'far apart' 'races' would produce what they called biological 'disharmonies.' Mendel's research was also used to buttress the contention that heredity differences between socially defined racial groups are beyond environmental influences and thus irreversible. This notion ran counter to Darwin's and many nineteenth century scholars' contention that the environment played a prominent role in genetic processes. As far as these scientists were concerned heredity and environment were irrevocably bound to one another. The debate between 'soft' heredity, which maintains that environment and heredity are intertwined, and 'hard' heredity, which asserts that inheritance is immune to environmental influences, was driven in large part by Charles Darwin's cousin, Francis Galton, who coined the phrase *nature versus nurture* to support the 'hard' perspective.[4] Contrary to Darwin, Galton argued that scientists can observe and calculate how biological traits pass from generation to generation without knowing the biological processes through which this occurs.[5] For Galton as long as scientists could observe and tabulate the transmission of traits from generation to generation, it was possible to control phenotypical outcomes.[6]

The idea that it is possible to directly observe and measure biological traits without knowing the molecular mechanics became known as *biometrics*. This Galtonian take on heredity and science dominated early genetic thinking in part due to Galton's protégé Karl Pearson's promotion of it in his influential book *The Grammar of Science,* and in the subsequent journal *Biometrika* that he founded in 1901.[7] In these publications Pearson maintained, like his mentor, that since objective creations are in principle unknowable, it is impossible to determine their 'real' causes and the best scientists can do is to use mathematics, in particular statistics, to explain scientific phenomena without actually understanding their underlying cause.[8] Hence if statistical correlations show inheritance works in a particular manner, the scientists' work is done. Since the statistical outcome (phenotype) is assumed to be the only measurable and explainable variable, there is no need to expound on why there is a positive or negative relationship between inheritance and other variables. This form of statistical thinking took the group "good of the race's" genes as its "central ideological value."[9] This is underscored by Galton's use of statistics to rank social groups' intellect, supposedly an inherited outcome, in his book *Hereditary Genius*. He uses this finding to convince Britain to establish public policies designed to assist "the more suitable races" triumph over "the less suitable races." To this end, Galton encouraged 'scientific' marital arrangements to breed intelligent children in order to improve human genetic traits through reproductive eugenics.

Because Galton believed 'white' Europeans like himself were far more intellectually advanced than other peoples, the design of his statistical methodology ensured he confirmed his assumption. In *Hereditary Genius,* Galton constructed a statistical grading scale that sorted each socially defined racial grouping into racial taxonomies to plot their range of intelligence. On this scale, he interpreted Greeks as having the highest intelligence; Brits and their descendants were two 'grades' below Greeks; Lowland Scotsmen and Northern Englishmen were a fraction below the average English; Africans were two 'grades' below the average English; and Australian Aborigines were last—one 'grade' below Africans.[10] Galton's racial and class inflected measurement scale ensured he found that 'inferior' socially defined racial groups and the poor were losing the evolutionary battle for existence. To stave off transmitting 'inferior' genetic material to 'intellectually superior' middle and upper class 'white' Europeans, Galton pushed for restrictions on human reproduction to curtail the birth of genetically 'inferior' individuals. Galton's concern with English intelligence emerged from his belief that the poor English classes' 'inferior' inheritance impeded Britain's imperialist project. As he put it: "the average standard ability of the present time should be raised" because societal advancement "call[ed] for more brains and mental stamina than the average of our race possess."[11] Since Galton embodied this commonsense mindset prior to creating his statistical measurement scale of intelligence, it was inevitable he would use value-laden categorization to rank socially defined racial groups in a linear manner based on immutable traits and natural ability.[12] With this biased statistical design Galton interpreted his field observations in Africa to suggest the continent's people were 'two grades' below the 'normal frequency distribution' of Anglo-Saxons' general mental ability. Galton portrayal of society as an accurate reflection of biology through statistical inference encouraged people to believe that social and economic inequalities between different groups—primarily 'races,' classes, and sexes—emanate from inherited, unchanging, inborn distinctions. Essentially, Galton's statistical analysis of human intelligence gave him and other biologists a 'scientific' way to legitimate racial, class, and gender prejudices and domination.

Initially Galtonian proponents ignored Mendelian theory because biometrics examined continuous (e.g., a complete range of variation in height or skin color) rather than discontinuous variation (e.g., number of distinct classes or categories, based on features like blood type that cannot be measured across a complete range). Nonetheless, both Mendelian and Galtonian ideas played significant roles in developing the ideological moorings of scientific racism. Though Galton's eugenic ideas developed slowly until the 1900s, due largely to their inability to show specifically how inheritance works, this changed when Mendel's laws of inheritance were rediscovered. The resuscitation of Mendel's heredity combined with the troubling belief that all or at least most human characteristics are genetically determined and passed down from parent to offspring to provide the basic elements for the rise of the eugenics movement.[13] The notion that heredity is immutable and therefore predictive of people's genetic 'health' coalesced with white supremacist meanings and frames legitimating 'white' dominance over marginalized socially defined racial groups. In short, ascension of genetic determinism at the turn of the twentieth century, particularly within the United States, validated the 'race' and class-based organization of society. One such example is how the concept of 'feeblemindedness' is applied. Feeblemindedness worked as a catch-all phrase that attributed mental retardation and any failure in scholastic performance, pauperism, lack of musical ability, degeneracy, alcoholism, as outcomes of 'defective' heredity.[14]

Charles Davenport, the first biologist to devote attention to the Mendelian character of human heredity, was the major force driving the growth of the eugenics movement in the United States. An associate professor at the University of Chicago, Davenport served on the editorial board of Pearson's *Biometrika* and used both statistical inference and Mendelian genetics in his effort to understand human heredity. The Brooklyn Institute of Arts and Sciences at Cold Spring Harbor, New York appointed Davenport director of its Biological Laboratory in 1890. During his tenure he persuaded the Carnegie Institute to set up a research center on evolution at Cold Spring Harbor. With Davenport as its director, the Station for Experimental Evolution opened in 1904 to investigate the confluence between theories of genetics, evolution, and cell biology. In 1910, with financial backing from railroad heiress Mary Harriman,

Davenport established the Eugenics Records Office (ERO) at Cold Spring Harbor to advance research on human heredity and to educate the public about eugenics.[15]

Despite the fact the theory of evolution and Mendelian genetics maintained that 'races' are differentiated by geographical alignment rather than biological essence, the latter misconception held sway due to the eugenic contention that heritable changes in phenotype (appearance) or gene expression signified the existence of biological 'races.' This claim enabled eugenics to put forth a pseudo-scientific means to rank order humans in racial categories according to ability, intelligence and moral character.[16] Such hierarchizing was needed in order for eugenicists to cement the view that social problems are irreversible inherited traits whose elimination required minimizing undesirable traits and maximizing desirable traits. In the early 1900s, concerns about social problems heightened with the massive influx of southern and eastern European immigration to United States which caused alarm among 'white' Anglo-Saxons who viewed them as a racial threat to 'white' Anglo-Saxon purity, power, and cultural hegemony.[17] This view was widely shared by 'whites.'

> Labor leaders worried about the new immigrants taking jobs from their traditional constituencies ['white' Anglos], and many ['white'] conservative Americans were concerned that the new immigrants were political radicals espousing Marxist ideas. Many ['white'] Americans worried that the immigrants were Jewish or Catholic, and thus unable [or unwilling] to assimilate into the traditionally ['white' Anglo-Saxon Protestant dominant culture of the United States].[18]

Given that eugenicists embodied the dominant ideologies of their day, they expressed the common belief that people are poor not due to inadequate educational and economic access but rather biological deficiencies in moral and educational capacity. Davenport's embodiment of this unsupportable scientific assumption led him to argue that intermixing with 'non-white' groups, whom he viewed as entirely different elementary species, would rapidly lead to the 'white' Euro-Americans becoming "darker in pigmentation, smaller in stature, more mercurial . . . more given to crimes of larceny, kidnapping, assault, murder, rape and sex-immorality."[19] He made this claim even though biological and anthropological research at the time indicated mating between members of different socially defined racial groups is biologically favorable and never detrimental.[20] As Dunn and Dobzhansky explained:

> Contrary to opinion vociferously expressed by some sincere but misguided people . . . a trend [toward 'race' fusion] is not biologically dangerous. Mixing of closely related 'races' may even lead to increased vigor. As for the most distantly separated 'races,' there is no basis to think that either biological stimulation or deterioration follows cross-

> ing. The widespread belief that human 'race' hybrids are inferior to both their parents and somehow constitutionally unbalanced must be counted among the superstitions.[21]

This supported Charles Darwin's earlier deduction that all human groups are members of the same species, as opposed to being separate species of humans.[22] The simple biological fact is that if humanity consisted of distinct 'races,' the resulting offspring from interbreeding among them would be sterile. Since intermixed human offspring are clearly not infertile, Dunn and Dobzhansky, like Darwin, correctly surmised our common ancestral origin. Despite this strong biological argument, eugenicists proffered a genetic essentialist-based interpretation that specified immigrants from Northern and Western Europe as possessing the 'superior' genes necessary to sustain the conscientiousness and love of learning and culture found among the native born 'white' population.[23]

It was a common sentiment during the nineteenth and first half of the twentieth century that certain racial and ethnic groups were superior in terms of moral intelligence and moral character. For instance, Germans as a rule were characterized as "thrifty, intelligent and honest," while Italians were seen as having a "tendency to personal violence."[24] Such racial characterizations operated as the foundational pillars for the U.S.'s Manifest Destiny doctrine which justified the genocide of Native Americans and the conquest of Puerto Rico and the Philippines based on the idea that that they were 'genetically inferior' and needed the protection and direction of 'genetically superior' 'white' Anglos.[25] But according to eugenic anthropologist Lothrop Stoddard, the immigration of these 'inferior' conquered groups to the United States was objectionable because their melding with the native born genetically 'inferior' population would constitute a threat to the reign of 'white' Anglo-American Empire. In Stoddard's words:

> If white civilization goes down, the white race is irretrievably ruined. It will be swamped by the triumphant colored races, who will obliterate the white man by elimination or absorption . . . if the present drift be not changed, we whites are all ultimately doomed. . . . [The] migrations of lower human types like those which have worked such havoc in the United States must be rigorously curtailed. Such migrations upset standards, sterilized the better stocks, increase low types, and compromise national futures more than war, revolutions or native deterioration.[26]

Davenport's and Stoddard's deftness at relating 'race' to undesirable traits generated enough popular support to turn their eugenic theorizing into public policies restricting immigration, enforcing involuntary sterilization and prohibiting interracial marriages. However, they did not spearhead the push for eugenic legislation. This task fell to Harry Laughlin, director of the Eugenics Record Office from 1910–1939. According to

Paul Lombardo, Laughlin was "among the most racist and anti-Semitic of early twentieth-century eugenicists."[27] Laughlin, like Davenport and Stoddard, was alarmed about the "great mass of defectives" within and immigrating to the United States.[28] Laughlin's deeply held belief that the first Nordic settlers to America, pruned by pioneer conditions, were the master 'race,' shaped his eugenic thinking and actions. Given this belief Laughlin tirelessly lobbied to prevent Jews fleeing Nazi Germany from immigrating to the United States.[29]

Prior to Germany's attempt to exterminate European Jewry, in 1922, Laughlin drafted a eugenics law template entitled, "Eugenical Sterilization in the United States," which was ostensibly designed to convince states to legalize sterilization of their "socially inadequate" populations, a designation which included

> . . . (1) feeble-minded; (2) insane (including the psychopathic); (3) criminalistics (including the delinquent and wayward); (4) epileptic; (5) inebriate (including drug-habituates); (6) diseased, (including the tuberculous, the syphilitic, the leprous, and others with chronic, infectious, and legally segregable diseases); (7) blind (including those with seriously impaired vision); deaf (including those with seriously impaired hearing); (9) deformed (including the crippled); and (10) dependent (including orphans, ne'er-do-wells, the homeless, tramps, and paupers).[30]

The draft was carefully crafted to pass constitutional muster to ensure states did not hesitate adopting it. Virginia used the draft's template to construct and pass a state sterilization law in 1924, which the U.S. Supreme Court ruled constitutional in its *Buck v. Bell* decision in 1927.[31] In this case, the Court legalized preserving economically privileged 'white' Euro-Americans' 'superior' heredity through compulsory sterilization. In the Court's ruling, Justice Oliver Wendell Holmes proffered the opinion that "three generations of imbeciles are enough" when referring to Carrie Buck, a rape victim the Court deemed socially unfit or feebleminded for having an illegitimate child.[32] Buck, her mother, and her daughter Vivian were ordered sterilized against their will to prevent the spread of idiocy. What is striking about this case is that neither Carrie, nor any members of her family, were mentally ill or possessed less than average intelligence.[33] In essence, because of Carrie Buck's mother's "immorality, prostitution, untruthfulness and syphilis," she and her daughter were involuntarily sterilized on the presumed basis they lacked social and moral status.[34] Doctor Albert Priddy, the prison warden overseeing the incarceration of Carrie and her mother for their 'crimes,' voiced this assumption in his Court testimony when he declared that the Buck family belonged to a "shiftless, ignorant, worthless class of anti-social whites of the South."[35] The Court's reliance on testimony and science grounded in erroneous facts, pseudo-scientific theory, and incorrect analogies assured that Buck would be involuntarily sterilized.[36] Contrary to genetic essentialist claims specifying Carrie Buck's daughter, Vivian, would be a degenerate, she attended Virginia's prestigious Venable School where she excelled as a member of the honor roll.[37] She did not become a third generation imbecile. Despite the Court's evidentiary fallaciousness, its ruling instigated the passage of numerous sterilization laws until the mid-1960s.

With regard to the eugenics movement, the United States set an example that other nations sought to emulate. Between 1928 and 1936, a number of European nations passed sterilization laws, beginning with Switzerland in 1928 and ending with Estonia in 1936.[38] German and American eugenicists, whose racial ideology was similar, regularly translated each other's literature, and the German movement was closely followed in the American eugenics press.[39] In 1936, Laughlin was offered an honorary degree from the University of Heidelberg in recognition of his work on behalf of the "science of racial cleansing." Laughlin gladly accepted the acknowledgement as he wrote, "not only as personal honor, but as evidence of the common understanding of German and American scientists of the nature of eugenics as research in the practical application of those fundamental biological and social principles which determine racial endowments and the racial health . . . of future generations."[40] Laughlin's words indicate that the American and German eugenic movements deeply shared "the identification of human beings as valuable, worthless, or of inferior value in supposedly hereditary terms."[41] Eugenics in both Germany and the United States was synonymous with purifying the 'white race' of 'low grade' and 'degenerate' groups. To facilitate the genetic cleansing of their nations' populations of those identified as carriers of unfit genes, eugenicists created a generic term: the genetically inferior.[42] This new label and its accompanying narrative gave rise to laws that sterilized individuals primarily because of their subordinate racial and economic status.

As the eugenics movement grew in popularity it developed a network of foundations, associations, journals, and academicians to generate scientific credibility for its ideology of heredity. For example, a year after receiving the honorary degree from Heidelberg, Laughlin established the Pioneer Fund foundation to support and publicize "heredity and eugenic" studies and "the problems of race betterment."[43] The Pioneer Fund supported the distribution of the pre-war 1930s Nazi Party's eugenic film *Erbkrank* ("Hereditary Defective" or "Hereditary Illness"), which was obtained by the foundation's financial benefactor William Draper, a reclusive heir to a textile fortune, from the predecessor to the Nazi Office of Racial Policy (*Rassenpolitisches Amt*).[44] Draper secretly met with Dr. C. Nash Herndon, a faculty member of Bowman Gray School of Medicine, at Wake Forest University in 1949. Though no record exists detailing what occurred during the meeting, after their encounter Herndon started playing a major role in the expansion of the compulsory sterilization program in North Carolina.[45] Herndon was an enthusiastic eugenicist whose 1943–1944 annual report for the Department of Medical Genetics described the eugenic sterilization effort in the county where his university was located by using language eerily reminiscent of Nazi experimentation.

> In September 1943, a project aimed at eugenic improvement of the population of Forsyth County was begun in co-operation with Dr. J. Roy Hege, Forsyth County Health Officer. This project consists of a gradual, but systematic effort to eliminate certain genetically unfit strains from the local population. About thirty operations for sterilization have been performed . . . "[46]

Since these involuntary sterilization procedures were not approved by the Eugenics Board of North Carolina, they were illegal.

EUGENICS PERSISTENCE

After World War II, popular support for eugenics gradually dropped due in large part to both general revulsion towards Nazi eugenic practices and geneticists' gradual realization that the theory of evolution and Mendelian genetics were right about human phenotype arising from complex interactions between genes and environment. Consequently, 'white' Euro-American eugenicists tried to distance themselves from their European/German counterparts to retain credibility with the public, but this proved impossible largely because their past correspondence with German eugenicists documented their enthusiastic support for the German eugenics regime.[47] In an attempt to evade the link to German eugenics, leaders of the movement began to repackage their message by recasting their desire for "better humans" in the broad terms of *improving public health*.[48] Eugenics successfully changed its racist façade, as systemic racism theory delineates, but not its racist purpose. Eugenicists ceased using language describing externally visible traits of 'race,' and began stressing the importance of molecular differences beneath the skin as markers of distinct racial divisions and diseased bodies.[49] In accordance with this change, the American Eugenic Society, which was established to promote eugenics, renamed itself in 1972 the Society for the Study of Social Biology while the journal of *Eugenics Quarterly* was retitled as *Social Biology* in 1969.

Eugenics proponent William Draper's funding support for anti-immigration and genetics research in the 1950s and 1960s by two government committees also ensured that eugenics endured.[50] The committee members included Representative Francis E. Walter (chair of the House Un-American Activities Committee and head of the Draper Immigration Committee), Henry E. Garrett (an educator known for his belief in the genetic inferiority of blacks), and segregationist Senator James O. Eastland of Mississippi, head of the Draper Genetics Committee.[51] Upon Draper's death in 1972 his estate bequeathed $1.4 million dollars to the Pioneer Fund to provide grants to "scholars" conducting research for "racial betterment." Past recipients of a Pioneer grant included William Shockley, who maintained that the higher rate of reproduction among the less intelligent has a dysgenic effect; Arthur Jensen, a leading proponent of black inferiority; and Audrey Shuey, author of *The Testing of Negro Intelligence,* which Arthur Jensen and Charles Murray used to formulate their racist theses.[52]

While eugenicists were busy reconstituting eugenics as genetics, technological advances in understanding genetic properties emerged making it possible to alter deoxyribonucleic acid (DNA). The eugenic assumptions underlying heritability resurfaced among many geneticists who believed technological advances about genetic structure provided them the means to eliminate undesirable genes and enhance 'desirable' genetic traits. These new developments in detecting and measuring human genetic variation also afforded geneticists the capacity to focus more on the processes by which human phenotypes take the form they do, rather than assuming that biology is immune to environmental influences, and thus static.[53] From this different emphasis a new concept termed 'population' emerged to describe groups breeding only within their group and sharing a specific genetic pattern. Since the term population recognized that individual within these categories often reproduced outside of them, populations are not considered to be discrete genetic entities in the natural world. The population concept apportions humans into *clines,* a geographical gradation of human genetic variation, rather than into biologically inauthentic 'races.'[54] So-called 'races' are permanent and static entities while populations are dynamic in their genetic compositions as a result of drift, mutation, migration and selection. Thus populations have biological reality while 'race' is a sociopolitical grouping.[55] As I discuss later in the book, population also operates as racial classification scheme, because 'race' is not simply replaced but reinvigorated as variable traits among groups rather than individual genetic differences.[56] This transformation of racial typology into a population does not relegate 'race' to the margins; it centers and validates it as an object of investigation and thinking.

The complex web of meanings about racialized 'other's' bodies is a part of scientists' cultural imagination that often goes unrecognized by genomicists because, as Marx put it, "Sie wissen das nicht, aber sie tun es" ("they do not know it, but they are doing it").[57] The very concept of racial ideology implies a kind of basic, constitutive naïveté: the misrecognition of its own presuppositions, of its own effective conditions, a distance, a divergence between so-called social reality and people's distorted representation and imaginary consciousness of it.[58] This is why 'naïve consciousness' should be submitted to critical-ideological analysis so naïve ideologically conscious genomicists can recognize ideology's conditioning effects on them. But it is not just a question of seeing things (that is, social reality) as they 'really are,' of throwing away the distorting spectacles of ideology; the main point is to grasp that human genome reality itself cannot reproduce itself without ideological mystification.[59] This façade does not simply hide the real state of genetics; racial ideological bias is written into its very essence. For example, Charles Davenport's endeavor to biologically engineer a society filled with people without disease and with longer healthier lives is operating now as medically oriented 'positive eugenics.'[60] Unlike their predecessors, geneticists today do not use 'negative eugenics' to restrict the mentally ill, poor, immigrants, and racialized 'others' from propagating. Rather the eugenic frame of "altruism and sentimental charity" is used to promote the less curative 'preventive' medical care for the genetically incomplete that seem to value individual health over sustaining the genetic health of supposedly 'superior' biological racial groups.[61] Leroy Hood, a pioneer of DNA sequencing, articulates subjective prevention as a means through which genomics is "mov[ing] from reactive mode (curing patients already sick) to a preventive mode (keeping people well)."[62] To prevent genetic problems, preventive measures are used to suppress genetically susceptible 'diseased' embryos and persons not through government coercion, but by passing eugenic decisions to the individual and/or relatives.[63] Such consent and compliance was unthinkable at the height of eugenics ideology. Although genomicists are not publicly pushing for eugenic public policies, the outcome tends to be the same given that scientific and social conditions are ripe for eugenic expression.

Eugenic ideology and practices are not relics of the past. Advances in genetic technology have given rise to 'preventive' genetic procedures like genetic screening that conflate the avoidance of disease with the prevention of lives deemed 'genetically imperfect.'[64] For example, fetuses with Down syndrome are regularly aborted in an attempt to prevent the passage of genes associated with the disorder to the next generation. This modern eugenic form is termed human genetic engineering. Human genetic engineering involves manipulating an individual's genetic makeup to alter his or her phenotype. Genetic engineering takes two forms—negative engineering, correcting genetic disorders and 'deficiencies,' and positive engineering, enhancing an individual's genetic composition.[65] Negative genetic engineering involves modifying or removing genes to prevent or correct genetic deficiencies through a procedure known as gene therapy. Gene therapy harnesses the powerful technology of recombinant DNA to correct an individual's diseased genes, which are then reintroduced into the person to replace his or her diseased genes.[66] But for the diseased cells to be replaced, self-renewing stem cells, or specialized cells that continuously divide to replenish all the cells with specific cell type, are required.[67] With negative genetic engineering genetic disorders are detected either prior to or during pregnancy. Prenatal diagnosis through amniocentesis (sampling embryonic fluid) or a simple blood test detects abnormalities in fetuses during the first trimester of pregnancy. Preimplantation genetic diagnosis (PGD) is another early form of 'preventive' diagnosis, in which embryos created in vitro are examined for genetic abnormalities. Only those free of susceptibility to disease are selected for implantation in the womb. PGD is used mainly in two broad indication groups. The first group are those individuals with high risk of having a child with a genetic disease, while the second group are individuals treated with in vitro fertilization, who might have a low genetic risk for abnormality but whose embryos are screened for chromosome aneuploidies[68] to enhance their chance for a successful pregnancy.[69] These two forms of negative genetic engineering operate as eugenics in that the information gathered from the screening is used to select embryos possessing 'desirable traits' for birth. Selecting offspring for their traits moves us towards a eugenic world in which people are valued more for their genotype than their inherent characteristics.[70]

Though genetic modification can potentially treat single gene mutations that cause genetic disease, susceptibility to cancer, and late onset disorders (e.g., Huntington's, Alzheimer's), it can also conceivably be utilized to 'improve' individuals in ways that may not serve a medical purpose. This kind of genetic modification is termed positive genetic engineering. Rather than reproducing 'desirable traits' via mating, individuals modify their DNA or the DNA of their offspring to enhance phenotype. The beginning of an enhancement/eugenic market is already visible in the United States, where Ivy League student donor eggs are priced 10 times higher than those of working-class women.[71] Prospective parents are now intervening not to mitigate the impact of their child's classed environment, but rather to manipulate their progeny's genes to increase the person's chances for economic success. So, in the genetic engineering context it is easier to alter a child's genetics than change the social conditions impacting their biology.[72] For instance, "one 'genetic condition' associated with reduced life prospects in many societies—the sex of the child—is easily shaped prior to birth using . . . PGD or ultrasound-plus-selective-termination."[73]

Because genetic screening is fraught with eugenic conceptions about what constitute a diseased body, genetic counselors and medical professionals' guidance is not merely technical advice. Their counseling and the individual's decision are shaped within a socially inflected genetic context that subtly persuades individuals to have 'fitter families' through genetic education that is framed as helping them, upon reflection, choose the best way forward for themselves and their families. Though most genetic counselors are committed to helping individuals discern the 'best' course of action for their families, David Helm and colleagues found that mothers who receive prenatal diagnosis of a genetic disability and choose to continue the pregnancy often encounter resistance from some medical professionals.[74] For example, a prospective parent who was informed there was a high probability her fetus had Down syndrome reported: "The genetic counselor treated me as though I couldn't accept this news, although I told her I could. She asked, 'What are you going to say to people when they ask you how you could bring a child like this into the world?'"[75] Through such concerned counseling, genetic counselors influence parental decision-making. Although many genetic counselors and physicians try hard to give information, not advice, their biased social dispositions regarding fetal 'fitness' are an unrecognized part of their thoughts and conversations with parents. Implicit mental aversion towards nonconformist phenotypes is a legacy of genetics' eugenic past that renders parents failing to act in accordance with eugenic preventive steps immoral.

The Continuous Geneticization of Social Problems

Though geneticists in the late twentieth century increasingly acknowledged genetics' eugenic past, most remain oblivious to how eugenic notions continue to manifest in their research and thinking. This is apparent in Daniel Koshland's (1989), editorial in *Science* magazine, for which he served as editor, asserting the genetic basis of social problems. He opined: "no group would benefit more from" the decoding of the genome "than the homeless, because diseases like depression and schizophrenia led them into poverty, and consequently homelessness" (1989, 189). Koshland's comments alarmed some genomic researchers, particularly prominent biologist S. E. Luria whose written response to Koshland's editorial insisted it "hints of a eugenic program targeted [at] 'the poor, the infirm and the underprivileged'" (Luria 1989, 873). He asked:

> are they to be transformed to (or perhaps altogether eliminated) by eugenic application of genetic technology? . . . The real danger today is the possible emergence of an establishment program . . . of beneficent eugenic intervention. Will the Nazi program to eradicate Jewish or otherwise 'inferior' genes be transformed here into a kinder gentler program to 'perfect' human individuals by 'correcting' their genomes in conformity, perhaps, to an ideal, 'white, Judeo-Christian, economically successful' genotype?[76]

Similar to eugenics, Koshland's "geneticization" of social problems assumes poverty, homelessness, crime and so forth are a function of genetic codes.[77] Geneticization is "the ongoing process by which priority is given to differences between individuals based on their DNA codes, with most disorders, behaviors, physiological variations . . . structured as, at least in part, hereditary."[78] As such, researchers immersed in the geneticization process pay scant attention to understanding how complex biological developments within and between individuals are shaped through interaction with various environments. For instance, researchers working to explain violent and criminal behavior usually invoke specific genes, hormones, and biochemical regulators to explain these social phenomena. This focus on biological differences as explanation for criminal behavior gained sway in the 1960s when research studies of penal populations suggested that males with an extra Y chromosome were prone to violence.[79] Despite the attempt to tie criminal and violent behavior to the presence of the extra Y chromosome in males (XY), later research found no direct relation between individuals with this chromosomal abnormality and aggressive behavior.[80] Yet, the search for the biological basis for violent and impulsive behavior continues. For example, a recent Dutch study[81] showed a very small correlation establishing the linkage between the desire to engage violent media in children ages five through nine and gene variation. The correlation explained less than 1 percent of the variance in individual behavior and did not establish a strong correlation between inheritance and behavior as suggested by the researchers who claim a strong linkage between genotype and violent media. Despite the fact the correlation percentage was insignificant, the authors argue that it is still important to pursue research in the area because "developments in genetics are currently revolutionary, and the explanatory power of genetic predispositions is expected to increase significantly in the coming years."[82] What this dubious and problematic contention implies is a vague acknowledgment that the relationship between genes and complex molecules like neurotransmitters is very complicated. However, subtle admission has not stopped genomicists from seeking the genetic causes of violence and criminality.

Neurotransmitters are thought to interact with regions of the brain involved in the expression of aggressive behavior. They consist of 5-hydroxtriyptamine receptors (5-HT for short, involved with the neurotransmitter serotonin): neurotransmitter and neurohormone dapmine, the metabolic enzyme monoamine oxidase A (MAOA), and a variety of steroid hormones (testosterone, androgens, and estrogen). The gene that codes for MAOA is presumed to be a central player in the manifestation of violence and criminality. Although studies examining the link between violence and the neuro-transporter gene serotonin—which transmits nerve signals in the brain to regulate sleep, sexual behavior, appetite, and impulsivity—acknowledge that environmental factors influence serotonin production, they ignore these elements in their quest to link the serotonin-transporter gene to violent behavior.[83] The MAOA gene which is found on the X chromosome interacts with 5-HT, regulating their release and breakdown so once they do what they are supposed to do they do not build up or interact with other receptors, causing communication problems between parts of the brain. In general, low 5-HT levels are associated with higher levels of aggressive or impulsive behavior and high 5-HT levels indicate low levels of aggression. When genomicists focus exclusively on biological mechanisms, they are inclined to downplay environmental effects in facilitating violence and categorize them as negligible because they view these effects as functioning primarily to bring out the inherited violent trait. This logic is apparent in a study by Caspi and his collaborators[84] exploring the significance of the interaction between MAOA and the experience of abuse in childhood (an environmental variable) for predicting violent behavior. They contend, when there is low activity of the MAOA gene, neurotransmitters in the brain (norepinephrine, serotonin and dopamine) are not properly metabolized and thus affecting mood. The stress of the childhood abuse brings out the phenotypical expression of the MAOA gene as violence. Though Capsi and colleagues[85] claim that their study includes an examination of the interplay between MAOA and the environment in producing violent behavior, their emphasis is really on the genetic etiology of violence.

The prominence of genetics over environment in conjunction with racist thinking and ideas incline violence researchers to place more correlative weight on the presence of two copies of the MAOA gene, found more commonly in African American males, in producing violent behavior.[86] Reti and colleagues' examination of the 2-repeat MAOA alleles determined that 0.5 percent of 'white' Euro-Americans possessed 2-repeat MAOA genes while 4.7 percent of African Americans do—almost tenfold difference.[87] Researchers interpret this small percentage in genetic variation as the reason African Americans are far more likely to engage in violent crime per capita as 'white' Euro-Americans hence 'explaining' their disproportionate incarceration rate. Their conclusion mirrors an earlier 2010 study of African American men and five gene loci suggesting these genes are predictive of violence and criminality more so than detailed measures of these men's environmental contexts (i.e., familial dynamics).[88] These claims reinforce the faulty notion that all African American males are predisposed to violence and criminality, though most black men do not carry the MAOA gene. African American men merely possess this genetic variation at a higher rate than 'white' Euro-American men. Moreover, carrying the MAOA gene does not necessarily lead to its phenotypical expression. This fact is easily substantiated by a cursory examination of arrest records overwhelmingly showing that the majority of African American males are in prison for nonviolent rather than violent offenses.

Despite the fact there is no evidence for any one-to-one genetic controls, nor is there evidence for certain molecules or systems in the body predetermining violent behavior, researchers continue their effort to decouple the effects of genes and environment in causing violence and criminal behavior.[89] So, why does the search for a biological basis of violent behavior persist? The reasons are many and confounding, ranging from the researchers' worldview to their need for public and private monies to fund their research. The latter funding source is directed at generating markets (e.g., genetic testing and pharmaceutical sales) to profit from genomic researchers' outputs/innovations. Current research exploring genes and violence also reveal a desire to link criminality to a distinct biological subspecies. This longing is a cultural force that is illustrated by the transformation of society's definition of a criminal offender from someone who has done bad things (morally guilty conduct) to someone who is defective thus beyond rehabilitation.[90] Since violence and criminality is located in genetically 'flawed' people, researchers do not consider how such behavior results from varying sociopolitical environments acting on biology. Though attempts to link inherited traits to criminal behavior have never held up in research, many researchers continue to argue that genes incline some socially defined racial groups to be violent. Commonsense racist notions about African Americans' natural aggressive tendencies ignore the complexity of human biology, psychology, history, and society. These ideas downplay the myriad ways aggression is initiated and maintained, and oversimplifies what is meant by, and understood as, human aggressive behavior. And, most dangerously, it enables a kind of inevitability in our communal sense of aggression and society.[91] Nobody is destined to be a criminal or violent. Certain powerful environmental and policy factors offer a more evidentiary plausible reason for differing incarceration rate among socially defined racial groups. The concern with genetic inheritance shaping marginalized socially defined groups' behavior reveals the extent to which deeply ingrained discursive racial constructions of the past and present about genetic normativity and differences remains salient in many researchers' imaginations.

HUMAN CLASSIFICATIONS AS RACIAL IDEOLOGY

In spite of results from a towering body of genetic and sociological evidence that is cumulative (observations over time), comprehensive (multiple levels of inquiry), and convergent (from a variety of sources, places, and disciplines) substantiating 'race' has no genetic basis, 'race' as genetic populations persists in biology. As I mentioned earlier, the equation of population to 'race' is not inadvertent given biology's definition of the term 'race' as a distinct genetic breeding group.[92] In other words, the population concept frames 'races' as variable traits that describe group rather than individual genetic differences to examine how the genetics of particular traits pass between generations within social groups.[93] As such, the population concept assumes a person is *probably/likely/maybe* genetically a certain way because *most/many/some* individuals belonging to his/her group are that way.[94] In chapter 3, I will clarify how these largely unexamined and unconscious assumptions embedded in population discourse drive genomic efforts to molecularize 'race' in disease and behavior in a way that pushes consideration of all other contributors to variation into the background.[95]

Genomics' need to generalize about these populations, as I discuss later in the book, can also serve as conduit for racial ideological expression. Since generalizations are broad statements or ideas about a group or thing, individuality is lost in average.[96] When genomicists do not account for this in their thinking and research their assumptions about genomic processes in groups serve to reify and strengthen abstract social types, such as 'races.' For example, in a series of studies U.S. genomicists purportedly found that self-identified 'race' corresponds to biological 'races' within the human species. This erroneous claim resulted from their questionable sampling and sorting of people into racial groups which they assumed were genetically similar. The researchers sampled genetic data only from people residing in the United States (who in no way represent global genetic diversity). Since the sampling group represented genetic material from people from the edges of the continents it gave the impression of an underlying biological 'race' structure within the human species when in fact there is none. In a word, researchers made racial typological assumptions about the nature of population genetic structure on the basis of a limited range of human genetic variance.[97] In their critique of the U.S. studies claiming continental genetic clustering that mimic 'race,' Serre and Pääbo argued:

> [T]here is a great tendency in the literature to use a few populations from the extremes of continental landmasses . . . to make worldwide inferences about substructures in the human gene pool. In fact, because human genetic diversity tends to be distributed clinally [continuously], it is especially problematic to sample the extremes of continents because this will create the impression of sharp discontinuities in the distribution of genetic variants.
>
> In this regard, it is worth noting that the colonization history of the United States has resulted in a "sampling" of the human population made up largely of people from western Europe, western Africa, and Southeast Asia. Thus, studies in which individuals from Europe, sub-Saharan Africa, and Southeast Asia are used . . . might be an adequate description of the major components of the U.S. . . . However, it would be incorrect to conclude that such studies [are] generaliz[able] to subdivisions of the human gene pool . . . worldwide. . . .[98]
>
> Studies of this sort with their narrowly constructed and racially inflected sampling method create racial groups based on *a priori* judgments that they exist in reality, thus they also exist in their data.

We are social beings who shape not only our environment but also, during the act of creating our social surroundings, our biology. My central point in this book is that systemic racism and its cultural components—whiteness and 'race'—in genomics are multifaceted and multivalent. Genomics is inextricably linked to racism and its ideologies. Given this it is simply impossible for either the 'race' concept or genomics to construct one without being implicated in the construction of the other. To claim this mutuality is nonexistent is a facetious act of bad faith that impedes genomicists' ability to recognize and account for racial ideology and practices within genomics. As the remaining chapters reveal, because genomic researchers exist within racialized history, the knowledge they possess is freighted to it and thus the tools they use to make sense of the human genome are products of that history as well. According to literary giant James Baldwin this occurs because

> History, as nearly no one seems to know, is not merely something to be read. And it does not refer merely, or even principally, to the past. On the contrary, the great force of history comes from the fact that we carry it within us, are unconsciously controlled by it in many ways, and history is literally present in all we do. It could scarcely be otherwise, since it is to history that we owe our frames of reference, our identities, our aspirations. And it is with great pain and terror that one begins to realize this. In great pain and terror one begins to assess the history which has placed one where one is and formed one's point of view.[99]

NOTES

1. Fullwiley, "The Molecularization of Race: Institutionalizing Human Differences in Pharmacognetics Practice."
2. Jackson, Jr. and Weidman, *Race, Racism, and Science: Social Impact and Interaction*, 103.
3. Scientific racism is the deliberate attempt to justify and protect a white supremacy using science as a source of authoritative knowledge.
4. Jackson, Jr. and Weidman, *Race, Racism, and Science: Social Impact and Interaction*, 99.
5. Jackson, Jr. and Weidman, *Race, Racism, and Science: Social Impact and Interaction*, 101.
6. Jackson, Jr. and Weidman, *Race, Racism, and Science: Social Impact and Interaction*, 101.
7. Pearson, *The Grammar of Science*.
8. Jackson, Jr. and Weidman, *Race, Racism, and Science: Social Impact and Interaction*, 101.
9. Rose, "The Politics of Life Itself."
10. Galton, *Hereditary Genius*.
11. Jackson, Jr. and Weidman, *Race, Racism, and Science: Social Impact and Interaction*, 102.
12. Simonton, "Francis Galton's Hereditary Genius: Its Place in the History and Psychology of Science."
13. Garver and Garver, "Eugenics: Past, Present, and the Future."
14. Jackson, Jr. and Weidman, *Race, Racism, and Science: Social Impact and Interaction*, 114.
15. Jackson, Jr. and Weidman, *Race, Racism, and Science: Social Impact and Interaction*.
16. Roberts, *Fatal Invention: How Science, Politics and Big Business Re-create Race in the Twenty-first Century*, 39.
17. Garver and Garver, "Eugenics: Past, Present, and the Future."

18. Jackson, Jr. and Weidman, *Race, Racism, and Science: Social Impact and Interaction,* 115.
19. Pickens, *Eugenics and the Progressives;* Kevles, "Eugenics and the Human Genome Project."
20. Provine, "Geneticists and the Biology of Race Crossing."
21. Dunn and Dobzhansky, *Heredity, Race, and Society,* 114.
22. Graves, *The Race Myth: Why We Pretend Race Exists in America,* 4.
23. Kevles, *In the Name of Eugenics: Genetics and the Uses of Human Heredity,* 46.
24. Kevles, *In the Name of Eugenics: Genetics and the Uses of Human Heredity,* 46.
25. Burgdorf and Burdgorf, "The Wicked Witch is Almost Dead: *Buck v. Bell* and the Sterilization of Handicapped Persons."
26. Stoddard, *The Rising Tide of Color against White World Supremacy.*
27. Lombardo, "'The American Breed': Nazi Eugenics and the Origins of the Pioneer Fund," 746, 756.
28. Lombardo, "'The American Breed': Nazi Eugenics and the Origins of the Pioneer Fund," 746.
29. Lombardo, "'The American Breed': Nazi Eugenics and the Origins of the Pioneer Fund," 756.
30. Laughlin, *Eugenical Sterilization in the United States;* Aultman, "Eugenomics: Eugenics and Ethics in the 21st Century"; Wellerstein, "Harry Laughlin's 'Model Eugenical Sterilization Law.'"
31. Wellerstein, "Harry Laughlin's 'Model Eugenical Sterilization Law.'"
32. Aultman, "Eugenomics: Eugenics and Ethics in the 21st Century."
33. Aultman,"Eugenomics: Eugenics and Ethics in the 21st Century."
34. Aultman, "Eugenomics: Eugenics and Ethics in the 21st Century."
35. Aultman, "Eugenomics: Eugenics and Ethics in the 21st Century."
36. Burgdorf and Burdgorf, "The Wicked Witch is Almost Dead: *Buck v. Bell* and the Sterilization of Handicapped Persons."
37. Lombardo. "A Third Generation of Imbeciles?" *DNA Learning Center Cold Spring Harbor Laboratory.*
38. Mehler, "Eliminating the Inferior: American and Nazi Sterilization Programs."
39. Mehler, "Eliminating the Inferior: American and Nazi Sterilization Programs."
40. Mehler, "Eliminating the Inferior: American and Nazi Sterilization Programs," 14.
41. Mehler, "Eliminating the Inferior: American and Nazi Sterilization Programs," 14.
42. Mehler, "Eliminating the Inferior: American and Nazi Sterilization Programs."
43. Lombardo, "'The American Breed': Nazi Eugenics and the Origins of the Pioneer Fund,"; Mehler, "The New Eugenics: Academic Racism in the U.S. Today."
44. Tucker, *The Funding of Scientific Racism: Wickliffe Draper and the Pioneer Fund.*
45. Begos, "Benefactor With a Racist Bent: Wealthy Recluse Apparently Liked the Looks and Potential of Bowman Gray's New Medical-Genetics Department."
46. Deaver, "Forsyth in the Forefront."
47. Garver and Garver, "Eugenics: Past, Present, and the Future."
48. Reardon, "Finding Oprah's Roots, Losing the World: Beyond the Liberal Anti-Racist Genome."
49. Roberts. *Fatal Inventions: How Science, Politics, and Big Business Re-create Race in the Twenty-first Century.*
50. Lichtenstein, "Fund Backs Controversial Study of 'Racial Betterment'"; the anti-immigration movement was deeply rooted in eugenic ideology.
51. Lichtenstein, "Fund Backs Controversial Study of 'Racial Betterment.'"
52. Mehler, "The New Eugenics: Academic Racism in the U.S. Today."
53. Stephan, *The Idea of Race in Science: Great Britain 1800–1960.*
54. Gannett, "Racism and Human Genome Diversity Research: The Ethical Limits of 'Population Thinking.'"

55. Gannett, "Racism and Human Genome Diversity Research: The Ethical Limits of 'Population Thinking.'"
56. Gannett, "The Biological Reification of Race," 323–345.
57. Žižek, *The Sublime Object of Ideology*, 28.
58. Žižek, *The Sublime Object of Ideology*, 28.
59. Žižek, *The Sublime Object of Ideology*, 28.
60. Comfort, *The Science of Human Perfection: How Genes Became the Heart of American Medicine.*
61. Abu El-Haj, "The Genetic Reinscription of Race."
62. Comfort, "The Eugenic Impulse."
63. Sparrow, "A Not-So-New Eugenics: Harris and Savulescu on Human Enhancement."
64. Buchanan et al., *From Chance to Choice: Genetics and Justice.*
65. Laura Hix, "Modern Eugenics: Building a Person?"
66. Laura Hix, "Modern Eugenics: Building a Person?"
67. Laura Hix, "Modern Eugenics: Building a Person?"
68. Chromosome aneuploidies is a chromosome abnormality where there is an extra or missing chromosome that results in genetic disorders including human birth defects.
69. Sermon et al., "Preimplantation Genetic Diagnosis."
70. Robertson, "Extending Preimplantation Genetic Diagnosis: The Ethical Debate."
71. King et al., "Eugenics fear over gene modification."
72. Sparrow, "A Not-So-New Eugenics: Harris and Savulescu on Human Enhancement."
73. Sparrow, "A Not-So-New Eugenics: Harris and Savulescu on Human Enhancement."
74. Helm et al., "Prenatal Diagnosis of Down Syndrome: Mother's Reflection on Supports Needed from Diagnosis to Birth."
75. Parens and Asch, "Disability Rights Critique of Prenatal Genetic Testing: Reflections and Recommendations."
76. Luria, "Human Genome Program," 873.
77. Lippman, "Prenatal Genetic Testing and Screening: Constructing Needs and Reinforcing Inequities."
78. Lippman, "Prenatal Genetic Testing and Geneticization: Mother Matters for All," 176.
79. Jacobs et al., "Aggressive Behavior, Mental Subnormality and the XYY Male."
80. Kim-Cohen et al., "MAOA, Maltreatment, and Gene-Environment Interaction in Predicting Children's Mental Health.: New Evidence and a Meta-analysis."
81. Nikkelen et al., "Media Violence and Children ADHD-Related Behaviors: A Genetic Susceptibility Perspective."
82. Nikkelen et al., "Media Violence and Children ADHD-Related Behaviors: A Genetic Susceptibility Perspective."
83. Williams, "Violence, Genes, and Prejudice."
84. Caspi et al., "Role of Genotype in the Cycle of Violence in Maltreated Children."
85. Caspi et al., "Role of Genotype in the Cycle of Violence in Maltreated Children."
86. Widom and Brzustowicz, "MAOA and the 'Cycle of Violence': Childhood Abuse and Neglect, MAOA Genotype, and Risk for Violent and Antisocial Behavior"; Rosenberg et al, "The Association of DNA Sequence Variation at the MAOA Genetic Locus with Quantitative Behavioral Traits in Normal Males."
87. Reit et al., "Monoamine Oxidase A Regulates Antisocial Personality in Whites with No History of Physical Abuse."
88. Beaver et al., "Genetic Risk, Parent-child Relations, and Antisocial Phenotypes in a Sample of African-American Males."
89. Fuentes, "Is Aggression Genetic?"
90. Moran, "The Search for the Born Criminal and the Medical Control of Criminality."

91. Fuentes, "Is Aggression Genetic?"
92. Gannett, "The Biological Reification of Race."
93. Gannett, "The Biological Reification of Race," 327.
94. Gannett, "Racism and Human Genome Diversity Research: The Ethical Limits of 'Population Thinking,'" S490.
95. Lock, "The Tempering of Medical Anthropology: Troubling Natural Categories."
96. Kattmann, "Race, Genes, and Racism."
97. Graves, "Evolutionary versus Racial Medicine," 147.
98. Serre and Pääbo, "Evidence for Gradients of Human Genetic Diversity Within and Among Continents."
99. Baldwin, "White Man's Guilt," 47–48.

TWO

Socialized Interpreters

Sociological examinations of racial thought in genomics eschew explaining it as only an individual creation and instead investigate its collective creation and internalization. I stress thought as a social creation because sociology is grounded in the basic principle that social facts—beliefs, tendencies and practices of the group taken collectively (coercively or willingly)—as realities external to the individual are crucial in shaping human action and thinking. In other words, individual actions and beliefs are not independent creations but rather constructions of enduring social patterns (i.e., norms, values, and relationships) that people internalize as their own in behavior and thought. As a consequence, writer Junot Diaz correctly observes, "we live in a [racial] imaginary where [most people] cannot conceptualize [marginalized groups] as fully human. What's really important is how this shit resides in us, how this just lives in us, man, even if we're the good guy—it should give a motherfucker pause."[1] Given this, genomicists' interpretations of genetic reality are necessarily products of their socialized understandings and thinking. In this chapter, I consider how the 'race' concept as an ideological component of systemic racism operates as an omnipresent ordinary narrative in genomic thought. By approaching racial ideology as accepted thought, we can begin to discern how many genomicists come to believe genes are the sole determinants of human biology thus disregarding the countless individual genetic variances shaped and reshaped by varying environmental contexts. As I argue in this chapter, racial reasoning as the hive mind of society inclines many genomicists to willfully surrender to the idea of biological 'race' as a *given*—in contradistinction to the constant questioning of accepted assumptions that science requires.

Scholars have long documented how scientific ideas and knowledge are major sites of racial contestation.[2] Numerous social scientists argue that contemporary knowledge continues to be racially coded and organized, and that racial thinking informs valued and dominant forms of knowledge.[3] Like these scholars, I am concerned with the consequences of internalized racial ideology on genomics and genomicists—with the self-deception and erroneous thinking it engenders, and the wasteful expense of time, money and energy expended to sustain the normative façade of systemic racism. Because systems of racial oppression are produced and reproduced in part through people's internalization of their cultures, this chapter assesses the social processes through which genomicists embody racial ideas that inform their genomic reasoning.

INTERNALIZING RACIAL CULTURE

Genomic knowledge is not just simply the result of scientific methods and reasoning; it is also a creation of racial hierarchical systems of domination, which generate and sustain racial meanings.[4] As a foundational cultural constituent that genomicists deeply imbibe, the racial ideology pervades every societal level—group relations, institutions, organizations, and power structures. Given its deep embeddedness in society, racial ideology is sustained concurrently at both the cultural and structural (material) levels in society.[5] At the cultural level 'race' as an ideology consists of values, norms, and shared patterns of seeing, thinking, and acting which genomicists consciously and unconsciously use in their construction of genomic reality. Genomic research's presentation of 'race' as biologically ordinary and normative often secures marginalized socially defined racial groups' oppression as when corporate media conveys its findings about molecular mechanisms of disease among socially defined racial groups as distinctive differences between groups. 'Race' typically comes across in such media reporting as an ordinary and necessary aspect of genomic discourse and practices as it did in a 2013 *Miami Herald* newspaper article about the importance of genetic anomalies in the ABCA7 gene for indicating a greater risk factor for African Americans to contract Alzheimer's disease. The piece noted that "while Alzheimer's occurs as frequently in blacks as other populations, there are important genetic differences among people of different races. . . . Identifying these differences may help researchers develop treatments and drugs that are more likely to be effective because they're tailor-made for an individual's genetic make-up."[6] In essence, the article implies that African Americans have a distinct genetic essence that genomicists can study to create new treatments and drugs tailor-made for an African American individual's genetic makeup.[7] Though the news report alludes to tailored individual therapies, it is grounded in the assumption of an individual's genetic sameness with their socially defined racial grouping. Racial ideology at the structural level involves racial thought that explicitly and/or implicitly justifies institutional actions, decisions, and policies that systematically privilege 'whites' in genomic research designs, thinking, and representations at the expense of marginalized socially defined racial groups. In contrast, my main concern in this chapter is to explore how genomicists' racial socialization cultivates racist thinking in genomics.

Genomicists acquire racial dispositions through the process of socialization whereby they take on racialized social norms and values as their own. During this process, genomicists as members of raced groups learn to think, feel, and act in ways conforming to the protocol of their racialized environment. Thus, socialization in racialized social settings engenders in genomicists specific messages and actions on personal and group identity, and social position related to racial status. From birth people acquire racial values and attitudes in subtle and sometimes unconscious ways that orient their own and others' existence as racialized individuals. Systemic racism, like all oppressions (sexism, classism, heteronormativity, etc.), is not simply a justification for exploiting marginalized 'others'; it is a dynamic belief system that is internalized and "performed from babyhood, [in a manner that] slip[s] from the conscious mind down deep into muscles and glands [where it] become[s] difficult to tear out."[8] This entrenched racist frame is directly implicated through meanings and perceptions, as sense-making that fashions how genomicists conceptualize the human genome. Consequently, genomicists' immersion in the habitus of racism makes it difficult for them to escape its influence on their thought.[9]

In racialized cultural contexts, genomicists experience what W. E. B. Du Bois terms a *double consciousness* where their social identities conflict with their sense of themselves as scientists.[10] Because of this, the struggle by genomicists with the consequences of seeing and situating themselves as individuals prescribed by racism results in cognitive dissonance. Although Du Bois's double consciousness concept refers to the duality confronting African Americans in the sense of experiencing a damaged self-concept as racialized 'others,' his theory is also relevant for understanding intra-racial socialization.[11] In the context of systemic racism all people manifest an uncertainty and variability about themselves, others, and social values.[12] Du Bois's concepts of the *racial veil* (racial lens) and *double-consciousness,* whose meaning and usage in his work are deeply intertwined, captures the conflictual nature of 'whites' experience in thought and ideals about themselves as racially color-blind individuals. The racial veil refers to the deep-seated demarcation of whiteness from blackness and the failure of 'whites' to lucidly comprehend themselves and others outside of the 'white' racial frame.

Writer Tim Wise expertly describes the racial veil's subconscious insidiousness in his description of its profound hold on his 'anti-racist' grandmother, whose Alzheimer's unleashed her suppressed socialized racial animus.

> [N]ear the end of my grandmother's life, as her body and mind began to shut down, this consciousness—the soundness of mind which had led her to fight the pressures to accept racism—began to vanish. Her awareness of who she truly was disappeared. And as this process unfolded, culminating in the dementia ward of a local nursing home, an amazing and disturbing thing happened: She began to refer to her mostly black nurses by the all-too-common term that forms the linguistic cornerstone of white America's racial thinking. The one Malcolm X said was the first word newcomers learned when they came to this country. Nigger. It was a word she would never have uttered from conscious thought, but one that remained locked away in her subconscious despite her best intentions and lifelong commitment to standing strong against racism. A word that would have made her ill even to think it. A word that would make her violent if she heard it said. A word that, for her to utter it herself, would make her another person altogether. But there it was, as ugly, bitter, and no doubt fluently expressed as it ever had been by her father (a member of the Ku Klux Klan).[13]

Though Wise's grandmother tried to undo her racist socialization through her work for racial justice, racialized thought remained deeply embedded in her psyche. Because racial ideology is imprinted subconsciously as obvious, it materializes in an institutional and normative way whereby individuals who maintain their distance towards racism are usually unaware of the fact that racist ideas already inform their thought and action.[14] The societal racial ideological enactment, not the depth of a person's inner convictions and desires, is what sustains racism in everyday life. As philosopher Slavoj Žižek observes: "the lesson is clear: an ideological identification exerts a true hold on us precisely when we maintain an awareness that we are not fully identical to it, that there is a rich human person beneath it: 'not all is ideology, beneath the ideological mask, I am also a human person' is the very form of ideology, of its 'practical efficiency.'"[15] Consequently, genomicists' attempt to distance themselves from racism does not simply lead to an objective understanding of biological processes; rather, it signals the existence of biological 'race' as normative in ways that fuels resistance to seeing it as a sociopolitical creation. Hence, the irrationality of racial thinking becomes embedded in the character of individuals to such an extent that though 'race' is not biologically tangible, people carry on as if it is real.[16]

Despite the overwhelming contention by genomicists that the application of scientific logic in genomics renders their work free of racial bias, racial assumptions and practices are commonplace. Jenny Reardon's research on the International HapMap (haplotype map) project's effort to record common patterns of human genetic variation found that it relied on the belief that meaningful genetic variability exists between racially defined social groups in order to assert that this variation is linked to human health and disease.[17] Specifically, Reardon argues the HapMap project researchers' reliance on categories of 'race' as biological proxies for genetic variation led to the reinscription of 'race' at the molecular level.[18] Though researchers explicitly designed the HapMap to mitigate racial bias, their reliance on geographical ancestry—African, European, Asian—to proximate the biological meaningfulness of the 0.1 percent difference between social groups ensured that it reflected familiar racial divisions. Moreover, as this interviewee explains, HapMap's limited sampling size of five geographical ancestral groups happened

> . . . because of funding limits . . . [which ensured the HapMap project did not take] representative samples from the entire Asia region, but from one village in China . . . one village in Africa . . . one town in Europe. Now, we already know that there are group differences in genetics based on geography. The criticism of the hap map project is that its small sampling . . . further reinforce[s] the notion of genetic difference based not on geography but on race.[19]

So instead of devising a more nuanced and comprehensive methodology for sampling, HapMap genomicists resorted to using social categories that easily mapped onto existing taxonomies of racial difference.[20] Reverting to racial grouping seemed reasonable to HapMap researchers because racial ideology is so profoundly imbued in their consciousness in ordinary ways. The embeddedness of racial thinking in genomicists' unconsciousness, as Du Bois informs us, makes us appreciate the fact that the persistence of the 'race' concept in genomics "[can] not be broken by a series of [intellectually] brilliant assaults." Defending the existence of bio-'race' "is not simply ignorance and ill will; . . . but also certain more powerful motives and [interests] less open to reason and appeal."[21] The work of racial socialization creates 'race' in genomics by providing researchers with racial cognitive frames that serve as the foundation for making sense of the human genome; and through its instilling of racialized culture which constitutes a veritable racial embodiment.[22]

Genomic thought, like any other mode of reasoning, must be problematized and critically engaged in terms of its interests, effects, and value as part of the broader racist environment. Genomicists must be attentive to how they function as racialized individuals and social agents so that they can confront the layered and complex social, political, economic, and cultural forces that shape and guide their genomic research. Inattentiveness to how racism animates genomic research can lead genomicists to create, reinforce, and reimagine human genetic variation not so much to challenge and overturn systemic racism but to solidify it as a 'commonsense' means for 'othering' the biological essence of socially defined marginalized groups. In this sense, racialized culture or meaning systems work not merely as a body of ideas but lived material practices—ways of thinking taking practical form—reproduced through the practices and production of major institutions like science.

SOCIALIZED RACIAL SILENCE

In racialized cultural contexts racial identity is a critical part of genomicists' individual and collective selves. The racial symbols, actions, and behaviors that genomicists acquire from family and community manifest as racialized conscious identities that shape their research and interpersonal relationships.[23] As an identity, 'race' involves self-understanding, social practices, and group beliefs that express what it means to be and live as raced individuals; as an ideology, 'race' systematically obscures how genomicists are racially marked in unconscious ways that are consciously opaque.[24] Consequently for 'white' genomicists, whiteness operates as "a form of social amnesia" or "epistemology of ignorance" that allows them to forget, to ignore, and avoid discussing their complicity in sustaining racism.[25]

This racism evasive maneuver is evident in the comments of this self-identified 'white' informant regarding the racial makeup of his childhood community: "It was pure white . . . very homogenous. It was all white and . . . Protestant. [Racial identity] was important but since it was so homogenous . . . [race] didn't actually [come up] in day-to-day life all the much." But as another self-identified 'white' interviewee explained, though racism was not discussed in majority 'white' communities, "I'm sure it was [important to people because] it came up every once in a while. We had a sort of typical non-diverse community. . . . I wasn't involved in a lot of discussions about racial identities, but it certainly came up every once in a while in conversation." Even though racist thought was always present at the surface of 'white' interviewees' communities, related racial discourse, practice, and policy were consistently silent and 'white' genomicists were socialized not to talk about them. This is one of the primary reasons 'white' respondents gave for evading racial discourse in their research and writing about genome variation among socially defined racial groups. As this interviewee explained:

> You . . . don't talk about this stuff and the reason is aside from there being a general unease about racial profiling, nobody wants to be accused of doing it. But deeper than that, at least in my case, my feeling is that people [should] be judged as individuals and you shouldn't prejudice people based on what you think is their social and ethnic background, to the extent that it is possible to do that. . . . So, I deliberately don't talk about these things. And if anybody starts talking about them, I generally tend to shut them up.

The respondent not only does not talk about racism, he vigorously pressures others not to talk about it, in effect demonstrating how 'white' genomicists actively silence 'race' talk.[26] Mica Pollock notes that such intolerance for speaking directly about 'race' is problematic in that

> [people], everyday, *are* reinforcing racial distinctions and racialized thinking by using race labels; but we are also reinforcing racial inequality by refusing to use them. By using race words carelessly and particularly by *deleting* race words, . . . both [researchers] and laypeople . . . help reproduce the very racial inequalities that plague us. It is thus crucial that we learn to navigate . . . race talk and colormuteness rather than be at their mercy.[27]

Not discussing racism and the effects of its associated ideologies ('race' and whiteness) in genomics tends to limit genomicists' ability to critically interrogate how racial ideology informs their research. It is only "when you name something, the words draw your attention to it, which makes you more likely to notice it as something significant. That's why most people have an immediate negative reaction to words like 'racism,' 'sexism,' or 'privilege.'"[28] Racial discourse is thus avoided not only to erase racial thought but also as a means to circumvent direct discussion of it, with the result that genomicists seldom perceived themselves and genomics as facilitators of systemic racism. In this respondent's words "deraced" discourse is common:

> . . . because race is a sensitive issue that we sometimes avoid addressing. The fact is that there are genetic differences between people, and those genetic differences are part of what makes up our physical body. Sometimes, the social construct of race makes people less capable of talking about differences in an honest way. There's been enough bad science, and misuse of science to try to imply one race is better than another. That, in a way, has tainted our ability to have an honest discourse about this natural variation. . . . We have a harder time talking about it, because of what has happened in the past around . . . race.

The phrase 'bad science,' which I will discuss more in chapter 5, is also often used to suggest that using racial discourse to talk directly about racism in genomics is indicative of racial bias, but instead it is in effect a means to mute discussion of the centrality of racial ideology in interpreting human genome variation. This ignoring or silencing suggests that racial thinking matters in genomics because 'race' is viewed as something to openly evade discussing among genomicists. Such evasiveness is conducive to the perpetuation of systemic racism because it serves to reinstantiate system-sustaining racial ideology.

Because racial silencing is prevalent in genomics, researchers centering racial ideology and racism within genomic discourse encounter colleagues, like the informant above, who seek to quiet their racial discourse. According to some interviewees, racial silencing of this sort is common in genome science because 'white' genomicists, who predominate the field, are unwilling to explore how their whiteness shapes their thinking and methodological designs. As this African American respondent explained:

> They don't have to [consider whiteness influence upon them] because the academy is always in that [white] framework. . . . [T]he academy [and the discipline] was founded in that environment. I have to say ['race' influences me and my work] all the time because I get looked at kind of strangely, or they say, "Well, why are you so adamant about . . . tak[ing] these positions?" . . . So, it's almost like I'm [judged by my 'white' colleagues as being] too emotional, or . . . un-objective.

SOCIALIZED COLOR-MUTENESS

Because most 'white' genomicists grew up never talking about racism with their families and neighbors it rarely occurs to them that their whiteness matters. In highly racialized societies, racial silence leaves whiteness unmarked and unchallenged in ways that people uncritically absorb as a mundane part of everyday life. As a consequence, most self-identified 'white' interviewees' only began to think about their whiteness during matriculation to post-secondary school where they encountered African Americans and other socially defined marginalized racial group members who encouraged them to consider the impact of their racial status on their lives. These reflections, however, generally sustained rather than challenged 'whites' own sense of whiteness, an outcome apparent in this self-identified 'white' respondent's comment:

> I just didn't think about [race]. Maybe this . . . was because I grew up in a place where race wasn't something I thought about. I had this roommate in graduate school who was bi-racial and that influenced me greatly, because she self-identified as black and I guess at that point I was thinking biologically, "Well if you're 50/50 . . ." And that was I think the first point at which I started to think, well it is about social identity to some.

Genomicists, like all people, are motivated to learn and conform to the broader racialized culture in order to function within it. In this cultural context, norms are acquired by genomicists from a broad range of sources—not just their immediate families.[29] Genomicists collect culturally acclimating information from peers, media, school, workplace, and many other important socializing agents that help them construct their own racial understandings. During childhood, genomicists learn which social categories are important in their cultural environment. They undoubtedly notice that people in their families and neighborhoods usually have the same physical characteristics. Therefore, genomicists, like others, assume and learn what they should avoid or dislike about people with a different phenotype than their own. The biases emerging from these interpretations "reflect both subtle and not so subtle messages about relative desirability of belonging to one social group as opposed to another."[30] Through racial socialization, whiteness is normalized and privileged. Consciously and unconsciously, genomicists internalized racial ideology as common sense. Such normative internalization is especially strong among 'whites' whom studies indicate, "rarely exhibit anything other than a pro-white bias."[31] This is in sharp contrast to marginalized racial 'others' whose social settings are permeated with talk about the negative impact of racism and whiteness on their daily lives.[32]

Because most 'whites' desire to disavow racism's influence in their lives, they believe the best way to prevent racism is not to talk about or notice it. This tactic is apparent in this response:

> [My town] was a very divided city. There was a black section and there was a Jewish section. And the section I live in was . . . white middle-class Protestant and Catholic. . . . I remember being aware of racial identity. I had a Jewish friend as a kid . . . who while his mom was driving us around would say, "Well here's where the Jewish neighborhood ends and here's where your neighborhood begins." I mean he was very up-front about it. I thought that was funny. I didn't personally socialize with black people until I went to college. . . . So my first experiences with black people were that they are separate.

Continuing:

> During my childhood I guess I thought in terms of how people [self-] identified or how my parents identified [them]. My friend, the one who made me think of Jewish people and people in school told me how to [racially identify people].

The respondent's friend's matter of fact talk about racial labels, ethnicity, and racial segregation unnerved him because he thought racial labeling signified racial bigotry.[33] Moreover, this respondent's comments marked marginalized socially defined racial individuals' discourse about racism as the problem, not his immersion in the racist contexts of his family and racially segregated neighborhood. The respondent's effort to make his whiteness invisible seems to be a product of his desire to deny how his whiteness implicates him in systemic racism.[34] Such evasiveness suggests that interviewees can think about racial identity or not, and that when they do think about racism they usually construct it as an individual not a structural or systemic creation. These constructions also indicate respondents valued avoiding being labeled a racial bigot more so than purging society of racial thinking and white supremacy, an inclination observed repeatedly in interviewees' attempts to embellish the racial diversity of their communities. For example, after this respondent initially claimed he was reared in "a town that was over 90 percent 'white,'" he later asserted the racial composition of his community was "probably 15 percent Asian, 5 percent African American." Such talk about 'race' intimates that 'white' interviewees believed it was something they did not have, but 'other people' did. This claim is dubious given that racism as an all-encompassing hegemonic force circumscribes how people perceive, narrate, and navigate reality. All genomicists are racial actors who are a part of a racial hierarchy that they do not transcend.

Though some interviewees rejected the 'race' concept due to its conceptual confusion and arbitrariness, 'race'-neutral approaches tend to lead most 'white' interviewees to downplay all racial differences in a manner that trivializes racial sociopolitical conditions' effect on the human genome. Such disregard of the 'race' concept also inclined respondents to normalize it within genomic practices and thinking. Given these deleterious outcomes, it is imperative that genomicists specify clearly that they reject racial essentialist claims regarding biological differences between socially defined racial groupings. Disavowing 'race' as biological distinction is by no means an easy feat given that genomicists live and perform racism. As such, racial ideology is created through genomicists' racialized social interactions where it enters their thought to delimit and facilitate genomic thinking.[35] This infusion of racial ideology into genomics directed racially socialized respondents to perceive the social conditions that racism produces not as creations of racial oppression, but rather as outcomes of the biological essence of marginalized socially defined racial groups. Consequently, 'race' as biological essentialism helps to sustain racial oppression. In this way, genomicists' racial position and thinking, like their class and gender locations,

> is constructed by a relation of persons to a materialized racist history that has constructed racially separated spaces, a racial division of labor, racist language and discourse, and so on. . . . [R]ace structures do not primarily name attributes of individuals or aspects of their identity but practico-inert necessities that condition their lives and with which they must deal.[36]

Yet, because most 'white' genomicists live racially homogenous lives, they mistakenly believe that racism is not a meaningful part of their lives.[37] This is far from reality given that racism's hegemony ensures that racial socialization is just as pronounced, as I mentioned earlier, among 'white' respondents reared in racially uniform communities as it is for those raised in racial heterogeneous neighborhoods. As this self-identified 'white' interviewee opined, whiteness in homogeneous communities is easier for individuals to embrace because "even though I grew up in a [racially] homogenous place, I watched TV, I'd go into the city where I would encounter more diverse folks. Basically diversity for me as a kid meant meeting blacks or African Americans in the [nearest city to my town]." Consequently, the respondent observed that before he learned to define 'race' via visual cues,

> and I still think that the way I understood [race] as a kid works pretty well. I think that race has its meanings in terms of practical day-to-day life and recognition. . . . I would make the argument that as you're walking down the street . . . just about every person . . . is on a moment-

> by-moment basis classifying with regard to race or ethnicity. And I don't know maybe there are people who don't do that. . . . [But] I would still define . . . I would still say, as a geneticist, a phenotype is anything that you assay. And I would say that race is assayed by each person. In other words, you identify yourself and others based on whatever criteria you judge to be relevant. . . . And I would say that most children learn how to racially phenotype as a very young age, sort of part of the classification of people walking down the street.

Such learned racial sorting taught respondents that phenotypical similarities imply group sameness and that racism—defined as overt, violent, and outdated—is simply an individual rather than structural phenomenon. Equipped with this myopic individualistic narrative, most 'white' respondents found it difficult to see themselves or their racially affiliate group as raced.[38] Given this limitation, informants preferred to merely note genetic differences between groups without talking about how their racial socialization informs their interpretation of these human differences. But to deal with racial thinking, genomicists must speak about 'race'—they cannot avoid it. Racism cannot simply be wished away by not talking about it. Naming the racial ideological and systemic racism dynamic in genomics is a very important and necessary first step for confronting the cultural obstruction of racial color muteness.

COLOR MUTENESS TO COLOR BLINDNESS

Because genomicists strive to be color-blind, most genomic respondents were reluctant to consider how color-blind reasoning masks 'race' as molecular differences. This concealment regularly occurs during comparisons of standard/healthy genomes [read: 'white' male biology] with nonconforming genomes, as this interviewee observed:

> [W]e really don't care why white men get prostate cancer. We want to know how to use *that as a comparator* to really understand why black men get prostate cancer and why their tendency to have aggressive diseases is so much statistically higher. . . . You could argue that that race has nothing whatsoever to do with susceptibility. It may have to do with other things. . . . But there is a fundamental underlying genetic clue that's there and I want to understand it.

While engaged in comparative analysis the sociopolitical character of 'race' is subsumed in biological contrast to suggest marginalized socially defined racial groups' genomes are distinctly different from those of 'whites.' Only a very small number of interviewees (mainly marginalized raced genomicists) recognized that using 'white' male genomes as a baseline necessarily implies marginalized socially defined groups' ('race' and gender) genomes deviate from "normal" ('white' male) genomes. These individuals, like this interviewee, noted how it was only through their work on African American susceptibility to disease that they "became cognizant about how much we know about Caucasian[39] genome and variation . . . and how largely ignorant we are about everybody else." However, their awareness of this did not necessarily motivate respondents to articulate a solution but rather to construct narratives blaming marginalized social groups' exclusion on their reluctance to donate samples. Such frames inclined respondents to assert that the practicality of their research trumped the span of their genetic data sampling:

> The design of my study would be based on what's feasible, unfortunately, what tends to be the most feasible in genetic research is . . . sticking with a straight Caucasian population, because there is so much information known about variation with the Caucasian population because there's just been a lot more people [having] their genome sequenced . . .

This suggests that the exclusion of a diverse set of racially defined social groupings in genomics is not merely product of these groups' refusal to participate in clinical studies. Respondents' responsibility narrative faulted racially defined groups, not systemic/structural impediments, for their exclusion in research. For oppressed racially defined groups, omission from clinical and biomedical research can cause them harm since they are not studied intensely enough to uncover clues about diseased biology. Including these racial groups' genetic samples is necessary both ethically and scientifically to detect how differing social and environmental factors impact on the human genome. The harmful consequence of excluding marginalized socially defined groups was apparent in earlier clinical studies of aspirin as a primary preventative for cardiovascular death in men. Though women were excluded from the aspirin study, the positive results of the therapy were assumed to be efficacious for them as well. However, aspirin therapy proved not to be an effective preventative for women.[40] In fact the aspirin regimen actually harmed them, demonstrating that it was not possible to effectively generalize the 'white' male research model to women and socially defined racial groups who may react differently to various drug therapies.

To address the exclusion problem, the National Institutes of Health in 1993 implemented the *Revitalization Act,* which set guidelines for inclusion and monitoring of women and socially defined racial groups in clinical trials.[41] The problem with the *Revitalization Act* is that it sustains the false idea that biological 'races' exist via its mandate requiring researchers to investigate "inter-population" rather than intra-individual genome variation. In other words, the *Revitalization Act* requires genomicists to use racial groupings as a proxy to infer intra-individual genome variation commonly referred to as *personalized genomic-based medicine.*[42] The racial configuration of this new genomic emphasis is apparent in the words of this respondent:

> I do think it is important to use the information that we have about the variants that are out there, and maybe it would associate with certain populations like the Jewish or African American population. I mean, I think it's important to recognize that there are some sequence differences . . . that can be useful for designer drugs. It's based on the sequence—it's genomic-based personal medicine.

On the face of it, the respondent's comments seem commendable, but sequence-based genomic medicine is grounded in the problematic assumption that racial categorization equates to genetic sameness. But as this informant argues, though there are some benefits to thinking about social group genomic variation given that people self-identifying or placed in the same racial categories do suffer from

> differences in incidences of diabetes, heart disease, cancer. . . . African Americans have a higher risk than almost any ethnic group, so there's clearly a correlation, but the big open question is, what, why that is? Is it access to care? Is it difference in socioeconomic status? Is it genetics? Is it a combination of all those things?

But, as I discuss later, respondents making multifactorial claims like the above interviewee tended to settle on genetic etiology to the exclusion of other variables because they are trained as geneticists. Their refusal to consider racism's impact on biology further facilitated interviewees' desire to analyze only genetic processes. This attempt to color-blur rather than color-focus also occurs because many respondents, like genomicist Keith Cheng, rightly feel they are out of their depth:

> While writing [out my research findings], I was carefully instructed by my colleagues to avoid using the term race, to the point that they encouraged me to deny its usability in discussions with the public. The reason is obvious: race has been used in a derogatory sense through human history. . . . On one hand, I cannot blame anyone for being sensitive about using this term. On the other hand, it seems to me that the laymen who use the term everyday would dismiss scientists as detached from reality if we refuse to discuss it.[43]

For this reason, although respondents recognize genomics cannot 'derace' reality, they also felt collegial pressure to feign racial color blindness when writing about and discussing human genome variation.

CAUCASIAN AS A 'RACE' EVASIVE IDENTITY

The depth to which genomicists, particularly 'white' researchers, engaged in racism avoidance is evident in interviewees' responses to the question: how do you racially identify yourself? Most white interviewees replied, Caucasian. Their use of this descriptor at first glance is puzzling given it is a geographical identifier referring to people from the Caucasus mountain region in Eurasia. Nevertheless, 'white' respondents identified themselves like these individuals: "I guess [I'm] just white or Caucasian" or "I'm a Caucasian American." Respondents using Caucasian as a racial identifier employed it interchangeably with 'white' or not at all to convey their belief that they are without 'race.' Specifically, their use of Caucasian insinuates they lack the capacity or resolve to reflect on their conformity to 'white' normativity. The mundane character of racial biases and prejudices make it difficult for respondents to detect it in their discourse. This is apparent in this interviewee's response to queries regarding her community's and her racial silence:

> Interviewee: You know, it's funny. Because, so interestingly, because I grew up in a farming community, of course, the Mexican-Americans were . . . and in fact, they may very well . . . been illegal immigrants for all I know. I just didn't pay attention at the time. And interestingly, because there were so few individuals (read: Mexican Americans), I don't think that . . . people just didn't care.
> Interviewer: Yeah. Yeah.
> Interviewee: . . . I mean, nobody thought about the fact that they were disadvantaged because they were doing something different than the rest of us.
> Interviewer: Well, what about the other, the inverse of that, though? Did you ever think about why was it that everybody was the same?
> Interviewee: Why was it that everybody was the same?
> Interviewer: Yeah.
> Interviewee: Did we think about that?
> Interviewer: Yeah.
> Interviewee: No, not really. We didn't ever consider that.

Notice that the respondent employs the discourse of sameness to dismiss racism's influence on her while accentuating whiteness as normative through the 'othering' of Mexican-Americans as foreign. Rhetorical machinations of this sort were common among respondents self-identifying as Caucasian.

The Caucasian self-conception provided 'white' respondents the means to dissociate from whiteness without depriving themselves of its fallacious racial assumptions regarding racial group members' biological similarity.[44] This imagined racial distancing encouraged many 'white' interviewees to believe it is possible to evade and erase whiteness in their thinking and research. This outlook is exemplified in the following exchange:

> Interviewer: In your opinion does your situatedness in a racialized society impact how you design a study? Is there concern about that?
> Interviewee: Well you know, I don't propose to be all knowing or whatever. My guess is that there probably are subconscious aspects to it, but I guess by virtue of definition . . . I'm not necessarily cognizant of the. . . . You know I feel like I'm reasonably *color-blind* to be honest with

> you. And I know that's tough, maybe incredible coming from a white person, but I feel like, as a scientist I'm a curious person by nature, right? So I don't tend to have my curiosity tempered by the fact that what I'm looking at might be only applicable to whites, blacks, Asians, what have you. And quite frankly, what's quite intriguing to me is when I think about the diversity of the world, is the diversity of disease as it affects different populations. And so, to me that's just all part of the mix.

The term Caucasian is deeply implicated in racism given that it was coined in 1775 by Johann Friedrich Blumenbach in his book *On the Natural Variety of Mankind* to connote the existence of human varieties/'races' diverging from the first human type, the Caucasoid.[45] According to Blumenbach, distinct human 'races' arose because Caucasians, the original humans, de-evolved into an African, American, and Asian variety—a contention that is the complete reverse of contemporary scientific evidence indicating *Homo sapiens* originated in the African continent.[46] Moreover, as one of Blumenbach's racial typologies, Caucasian was used by him to denote this socially defined human grouping as the possessor of an idealized physical appearance far superior to that of its supposedly de-evolved kindred. The logic that 'whites' physical appearance indicates advanced development is implicitly racist given that it equates socially constructed notions of beauty with *better* genes.

Though this is the case, Caucasian retains its cache among many 'white' informants as an unquestioned identifier and racial distinction frame. The term's continuous use is a telltale sign of respondents' willful ignorance of the power of systemic racism in genomics.[47] Although many respondents are distressingly uninformed about the racial biases from which 'Caucasian' emerged, obliviousness as an excuse for misuse of the term is unacceptable since it is incumbent upon genomicists to ensure that the term is defined accurately to assess whether or not it is indeed what they believe it to be—a distinct racial category. Despite the fact 'white' respondents see themselves being as fair-minded individuals whose thinking and research are impartial, they, like the average 'white' individual, lead segregated racial lives that distort their perceptions of marginalized socially defined racial groups' lived experiences and biology. This separation predisposed most 'white' interviewees to falsely assume marginalized socially defined racial group members share their outlook and experiences. Given the distorted lens created by the presumed universality of whiteness, there was little to no conscious thought among most 'white' respondents about how systemic racism and racialized culture delimits their experiences and thinking.

In view of this, respondents' use of Caucasian as an identifier signifies tolerance for the logical fallacies that it generates in their scientific thinking. For example, Caucasian as a concept of racial distinction spawns the circular argument that human beings are similar, so are members of the same species, but belong to several different 'races' that differ in important ways.[48] This reasoning ignores voluminous evidence "that phenotype is not equivalent to genotype—precisely because observed traits are a function of gene expression and not simply gene frequency."[49] Recognizing such fallacies in reasoning was difficult for most 'white' respondents given their involvement in the everyday logic of whiteness which evinces the view individuals may be genetically a certain way because their socially defined racial group is characterized that way. This logic is observable in noted genomicist Craig Venter's remarks:

> Like many people, I reach for my inhaler in smoggy conditions. Genetics contributes to this susceptibility and researchers have focused on certain family enzymes that help detoxify everything from carcinogens to pharmaceuticals. There is a gene that is associated with the ability to degrade environmental toxins; however, nearly half of the Caucasian population lacks the gene. . . . [S]o perhaps that is why I am more susceptible to environmental toxins.[50]

These comments clearly assume Caucasian means people identifying as such are biologically similar thus distinct. To use Caucasian as a color-blind identifier and scientific category in no way avoids racism; rather, it normalizes racist thinking by further embedding racial ideology into genomics' basic sense-making frames.

In sum, genomicists' claims of 'race-blindness' are highly unlikely given that racism, much like the air we breathe, is ubiquitous and normative. Genomicists are socialized into racial color blindness in ways that incline them to minimize and obscure how whiteness is marked and closely policed in their research and thinking to differentiate 'white' from racial 'others.' Moreover, racial ideology's influence in genomics is downplayed by most genomicists in order for them to discount their acceptance of racism as ordinary. Because genomicists live in a racial imaginary they cannot escape, they use color-blind ideological blinders to evade considering how racial thinking and practices enter the workings of genomics.

NOTES

1. Though the quote's original focus is on patriarchy I modified it to capture how it pertains to the racial imaginary described in the book. McDermott, "Lunch with Junot Díaz."

2. Smedley and Smedley, *Race in North America: Origins and Evolution of a Worldview*; Stephen Jay Gould, *The Mismeasure of Man.*

3. Zuberi and Bonilla-Silva, eds., *White Logic, White Methods: Racism and Methodology*; Feagin, *Systemic Racism: A Theory of Oppression*; Cazenave and Maddern, "Defending the White Race: White Male Faculty Opposition to a 'White Racism' Course"; Stanfield, II, "The 'Negro Problem' Within and Outside the Institutional Nexus of Pre-World I Sociology."

4. Reardon, *Race to the Finish: Identity and Governance in an Age of Genomics.*

5. Feagin and Elias, "Rethinking Racial Formation Theory: A Systemic-Racism Critique."
6. Chang, "UM Researchers Identify New Gene Associated with Alzheimer's in African Americans."
7. Chang, "UM Researchers Identify New Gene Associated with Alzheimer's in African Americans."
8. Smith, *Killers of the Dream*, 96.
9. Rutledge, "Socialization and Racism: The White Experience."
10. Du Bois, *The Souls of Black Folk*.
11. Rutledge, "Socialization and Racism: The White Experience."
12. Rutledge, "Socialization and Racism: The White Experience," 75.
13. Wise, "Famous Last Words: Exploring the Depth of Racist Conditioning."
14. Paraphrasing Žižek's words in *The Plague of Fantasies*, 5
15. Žižek, *The Plague of Fantasies*, 21.
16. Reich, *The Mass Psychology of Fascism*.
17. Reardon, *Race to the Finish: Identity and Governance in an Age of Genomics*; Jennifer A. Hamilton, "Revitalizing Difference in the HapMap: Race and Contemporary Genetic Variation Research."
18. Reardon, *Race to the Finish: Identity and Governance in an Age of Genomics*.
19. Jewish Currents, "Genetics, Race, and Socially Responsible Science: An Interview with Alan McGowan."
20. Hamilton, "Revitalizing Difference in the HapMap: Race and Contemporary Genetic Variation Research."
21. Du Bois, *Dusk of Dawn*, 6.
22. Paraphrase from Bourdieu and Wacquant, *An Invitation to Reflexive Sociology*.
23. Helms, "Introduction: Review of Racial Identity."
24. Helms, "Introduction: Review of Racial Identity."
25. McLaren, "Whiteness Is . . . The Struggle for Postcolonial Hybridity"; Charles Mills, *The Racial Contract*.
26. Pollock, *Colormute: Race Talk Dilemmas in an American School*.
27. Pollock, *Colormute: Race Talk Dilemmas in an American School*, 4.
28. Johnson, *Privilege, Power and Difference*, 11.
29. Hirschfeld, "Children's Developing Conceptions of Race."
30. Katz and Kofkin, "Race, Gender, and Young Children."
31. Katz and Kofkin, "Race, Gender, and Young Children," 62.
32. Lawrence A. Hirschfeld, "Children's Developing Conceptions of Race."
33. This also suggests that 'white' Euro-American racial socialization varies but they all still are socialized into whiteness.
34. McLaren, "Whiteness Is . . . The Struggle for Postcolonial Hybridity."
35. Satre, *Critique of Dialectical Reason*.
36. Young, "Gender as Seriality: Thinking about Women as a Social Collective."
37. Lewis, "'What Group?' Studying Whites and Whiteness in an Era of 'Colorblindness.'"
38. Bartoli et al., *Chasing Colorblindness: White Family Racial Socialization*.
39. The term Caucasian was often used by 'whites' as a self-identifying and a label identifying the genome of others of their ilk. The term is problematic because it postulates that the natives of the Caucasus exhibit idealized physical appearance so the Caucasus are believed to be the birthplace of mankind. The logic behind this idea—the assumption that 'whites' exhibit the best physical appearance—is implicitly racist. Additionally, we now know our species first appeared in Africa, so the biology is not sound either.
40. U.S. Preventive Services Task Force. "Aspirin for the Primary Prevention of Cardiovascular Events: Recommendation and Rationale."
41. Bliss asserts in *Race Decoded: The Genomic Fight for Social Justice*.
42. I will say more about this in chapter 5.
43. Glaser, "Investigator Profile: An Interview with Keith Cheng, M.D., Ph.D," 3.

44. Krieger, "Stormy Weather: Race, Gene Expression, and the Science of Health Disparities"; McDermott and Samson, "White Racial and Ethnic Identity in the United States"; Sleeter, "Advancing a White Discourse: A Response to Scheurich."
45. According to Thomas Gossett in *Race: The History of an Idea in America*, p. 39, the origin of 'Caucasian' was based upon a single skull in Blumenbach's collection.
46. Krieger, "Stormy Weather: Race, Gene Expression, and the Science of Health Disparities."
47. Krieger, "Stormy Weather: Race, Gene Expression, and the Science of Health Disparities."
48. Krieger, "Stormy Weather: Race, Gene Expression, and the Science of Health Disparities."
49. Krieger, "Stormy Weather: Race, Gene Expression, and the Science of Health Disparities."
50. Venter, "The Richard Dimbleby Lecture 2007: Dr. J Craig Venter—A DNA-Driven World."

THREE

Racialized Culture—Genomic Nexus

In the previous chapter I explored how genomicists' internalization of racial ideology inclines them to evade discussing the 'race' concept's impact on their genomic reasoning. I push this investigation further along in this chapter by showing how the imperceptible systemic character of racism and racist sensibilities trigger the racial signification of human genetic differences.[1] As a cultural expression, racial ideology plays an important role in constructing genomics, which in turn constitutes racialized culture. I use the concept of racialized culture to reveal "those instances where social relations [in genomics] . . . signif[y] human biological characteristics in such a way as to define and construct differentiated social collectives."[2] Racialized culture is a group/social process in which respondents partake in the production of racialized structures and institutions. In this chapter, I establish how respondents assign racial meaning to genetic material and processes based on how they are taught to perceive reality in racialized environments. Contrary to the claim that 'race'-blind methodologies ensure racial ideology is not a factor in genomic research, genomicists' aversion to the reality of 'race' suggests otherwise. Thus this chapter engages how racialized culture mediates genomicists' understandings and representations of human genetic variation.

Racial ideology and racism are social processes serving as key social influencers in teaching people that differences are important and racial identity matters. For this reason, the inscription of 'race' in biology is not a unidirectional but a bidirectional signification process where "ascribing real or alleged biological characteristics with meanings defining the Other necessarily entails defining one's Self by the same criterion."[3] So, 'white' interviewees self-identifying as "White Caucasian,[4] European derived . . . geneticist[s]" bring to genomics an understanding of themselves grounded in deeply internalized notions about 'race' and racial 'others.' Such thought is anchored in 'race'-neutral logic that enables respondents to avoid considering how the regime of white supremacy informs their genetic interpretations. Such dodging is unmistakable in this interviewee's remarks:

> I may start an argument with certain people, but I just don't feel it's [referring to 'race']. . . . I mean I understand that we sort of have this idea that there is no such thing as . . . research that doesn't come from a social context, it's sort of what we've all been taught, but I'm not feeling the social, class, or economic strata context of the work that I do. In the ways the outcomes are used might be, but . . . I don't sense that when I'm actually looking for a genetic variant that might contribute to a disease . . .

These comments are at odds with respondents' racialized identities which are not merely individual creations but rather signifiers generated and maintained by systemic racism structures. Yet, as I discussed in the previous chapter, 'white' interviewees are averse to considering how disciplined, systemic, and collective racialized activity structures genomics. Their reluctance stems in part from their conceptualization of racism as a highly conscious and deliberate set of acts infused with racial animus or intent. Racism is viewed as a form of personal bias rather than a structural occurrence.[5] For this reason writer Scott Woods argues

> that white people see racism as conscious hate, when racism is bigger than that. Racism is a complex system of social and political levers and pulleys set up generations ago to continue working on the behalf of whites at other people's expense, whether whites know/like it or not. . . . Racism is an insidious cultural disease. It is so insidious that it doesn't care if you are a white person who likes Black people; it's still going to find a way to infect how you deal with people who don't look like you. Yes, racism looks like hate, but hate is just one manifestation. Privilege is another. Access is another. Ignorance is another. Apathy is another. And so on. So while I agree with people who say no one is born racist, it remains a powerful system that we're immediately born into. It's like being born into air: you take it in as soon as you breathe. It's not a cold that you can get over. There is no anti-racist certification class. It's a set of socioeconomic traps and cultural values that are fired up every time we interact with the world. It is a thing you have to keep scooping out of the boat of your life to keep from drowning in it.[6]

With this internalized faulty subjective viewpoint informants find it difficult to conceptualize systemic, communal, and coordinated racial practices as collective exercises of group power persistently privileging 'white' genomes over those of marginalized socially defined racial groups. In addition, interviewees' destructured view of racism encourages them to erroneously believe they are not involved in the maintenance of systemic racism because they *lack* racial consciousness and hatred. This perspective fosters the illusion that whiteness is neutral in genomics. Because genomic interpretations and representations are presumed to be free of racial prejudices, when racially biased research is identified it is processed by interviewees as an exception rather than a rule in human genome inquiry.

Racialized culture functions through the production of formal and informal forms of discursive power that lay out the terms by which genomicists constitute themselves as subjects of their own conduct.[7] They are subjects in the sense of agents but also subjected to a form of racial power through which their actions and beliefs become legible as their own, when they are in fact group creations. Such racial subjectification materializes when genomicists accept racialized culture as their own, as we saw in chapter 2, through the process of socialization. Thus, in this chapter, I describe how racial ideology is absorbed consciously and subconsciously by genomicists as a guiding 'truth' that plays out in their conceptions of human genome variation.

GENOME VARIATION IN A 'RACE'-CONSCIOUS CULTURE

In a racialized social world colored by racialized subjectivities, definitions and actions, genomics, with its implied scientific precision and accuracy, is used in the same way that physical appearance was in the past to construct ideas and discourse suggesting 'race' is a biological fact. Though few genomicists believe there are sharply genetically differentiated geographical groupings among humans, many argue that although most human genetic variation does not fall along so-called racial lines, some genes are more prevalent within some socially defined racial groups than others. The prevalence of specific genetic variation in one group more than others is considered sufficient evidence for the biological basis of 'race.' This assumption is observable in biologist Armand Leroi's assertion that genetic data reveal 'races' exist:

> Certain skin colors tend to go with certain kinds of eyes, noses, skulls and bodies. When we glance at a stranger's face we use those associations to infer what continent, or even what country, he or his ancestors came from—and we usually get it right. To put it more abstractly, human physical variation is correlated; and correlations contain information. Genetic variants that aren't written on our faces, but that can be detected only in the genome, show similar correlations. It is these correlations that Dr. Lewontin seems to have ignored. In essence, he looked at one gene at a time and failed to see races. But if many—a few hundred—variable genes are considered simultaneously, then it is very easy to do so.[8]

Leroi's primary evidence for the biological basis of 'race,' however, derives from a methodologically flawed research that assumes individuals from the same socially defined racial groups are genetically similar.[9] Because, as I mention in preceding chapters, such studies collect genetic material based on the erroneous assumption that racialized continental groups (African, European, Asian, Native American) are genetically alike, they give the false impression that there is an underlying biological 'race' structure within the human species when in fact there is none. Consequently, genetic similarity is found among pre-labeled continental groups but no sharp genetic differences are evident among them. Studies of the genetic structure of human groups find some 'European' individuals are sometimes classified as 'Africans,' some 'East Asians' as 'Middle Eastern,' some 'Native Americans' as 'East Asians' and so on.[10] The point is that human groups do not fit well into socially constructed racial groupings, but genomicists' *a priori* judgment that biological 'races' exist in reality leads to the construction of constricted racially inflected sampling methods that ensure racial genetic similarities and differences are found. In short, the biological parsing of individuals into continental 'races' is problematic because it is usually inferred through genomicists' cultural beliefs about the sampling groups' 'geographic ancestry' and understandings of their genetically rooted phenotypic characteristics. Such notions are reflected in the last sentence of this respondent's comments:

> I certainly think disease has different prevalence by races. But I don't know if that's a genetic cause. I[t] . . . could . . . be environmental or socioeconomic, or any number of things that are quite different. So it's not necessarily genetic. . . . I don't see race as purely genetic construct. I see race . . . as some sort of history and ancestry of where you came from. . . . [But] to the extent that there is disease that differs among racial groups, I guess race can have some utility.

Similar to this interviewee, many respondents did not claim racial categories mapped directly onto inferred genetic racial groupings but suggested that biological groupings are important components for social group comparison instead of a basis for substantiating racial genetics.[11] As this interviewee put it:

> So if you measure 500,000 genetic markers . . . you . . . basically see who's more similar to who, some people are clustered closer together than other people, and those groups roughly overlap with what people colloquially call 'race,' but there's certainly no one-to-one correspondence, nor is one driving the other. . . . So . . . it turns out that there are many thousands of markers across the genome where the allele[12] frequencies are different across what we would call racial-ethnic groups. So African Americans have very different allele frequencies than European ancestry folks. And . . . then you have the differences in prostate cancer risk, or the differences in diabetes incidents. . . . There's no reason to think that those incident rates are actually due to differences in alleles, and genes. So in order to avoid that kind of bias, we, as a first

task, will often match people of self-reported ethnicity, self-reported race, so that the good first pass, but within those self-reported groups, there are some subtle, and some not-so-subtle gradations that are important to take account of. And given the last five years, we've gotten pretty good at measuring those subtle gradations, and it's pretty much standard to account for that now if you can afford to do the genotyping.

Though the respondent's contention is couched in a way to suggest allele differences among groups do not signify biological 'race,' the use of self-identified 'race'—which varies widely as an identity because people's racial affiliation differs depending on the context—as a sifter and comparator for detecting human genome variation unavoidably constructs genetic differences as racial distinction as evident in this interviewee's remarks:

> . . . I have attended talks where people describe the different causes for example kidney disease in African American and in Caucasians. And if you look at people who are on kidney dialysis they're very . . . ; the Caucasians are there because of complications with diabetes and the African Americans are there because of genetic variance to increase their risk to a disease called FSGS, it's Focal Segmental Glomerulosclerosis.
>
> And so it's just a very different set of risk factors for and a very different, they're really two different diseases that present to the same way. But when . . . [researchers] present this data, [they receive a] great deal of push back. Essentially people say that you should treat all, you know, all the patients medically the same way regardless of race, because they're all people and they should all react the same way.
>
> Yeah, I think . . . the fact is that different racial groups do have different genetic make ups and different genetic backgrounds. There's some diseases that are more common among people of African ancestry and some diseases that are more common among people of Caucasian ancestry. To ignore that as a geneticist or a medical professional is silly in the same way that you, that it would be silly to ignore, you know, smoking history if you're (inaudible) lung cancer.

The fundamental racialized character of genomicists' perceptions and methodologies makes it virtually impossible to definitively determine the link between racial taxonomies and the independent reality of genetic group.[13] Despite this genomicists continue their efforts to validate correspondence between 'social race' and biological groupings. Such work endures according to one respondent because:

> People believe what they do and scientific evidence does not sway the minds of individuals *including* scientists. I think that what science does and has done is in the mind of most people, [is to create a] sort of curious kind of coexistence . . . or accommodation. They believe what they believe and science is not going to change what they believe . . .

> though certain facts are to change . . . they're . . . going to accommodate it or coexist with it. But many, many people can live two quite separate lives. I find it extremely difficult, but many people do it . . .

Evidence for this informant's contention is found in the fact that despite repeated confirmation of evolutionary biologist Richard Lewontin's finding that human groups do not fall along racial lines given that of all genetic variation among humans (0.1 percent of genes vary among individuals)—85.4 percent occurs *within* geographical groups—genomicists continue to mine the remaining 14.6 percent for human racial differences by continent.[14] This singular focus on genetic differences among social groups encouraged respondents to ignore the significance of social determinants in fostering genetic variance as demonstrated by the partiality of this respondent:

> We believe there is a genetic basis [for prostate cancer]. Of course, it is not all genetic. There are also going to be lifestyle and environmental factors as well. But our findings . . . suggest that *a large fraction* of the disparity between African Americans and other populations *[is] due to genetic variation in this region* [emphasis mine].[15]

Interviewees' efforts to "find" what they insist are distinct genetic differences among groups serves to deemphasize how "it is not that different biological processes underlie disease formation in different groups, but that different life experience activates physiological processes common to all, but less provoked in some."[16] Although respondents framed their search for genetic differences as though they are not hunting for genetic racial variance, their discourse suggests this is precisely what they are doing. To skirt confronting the implicit racial character of their research, interviewees frequently used coded language such as populations rather than 'race.' This individual, for example, asserted: "I talk about genetic differences associated with certain populations that can lead to differences in disease states or responses to drugs . . . But I wouldn't put [the] label of race on it." This rhetorical sleight of hand is also voiced by another informant:

> I think there are populations that have been geographically isolated and then formed . . . their own [genetic] variants . . . within . . . their population because they were isolated from other populations that allowed these specific isolated populations to [develop] a common genetic makeup, but have their own particular variants they carry within populations. So yeah, I guess I do see race as sort of a genetic definition.

These comments intimate that the development of unique gene patterns as a result of environmental adaptation across social groups are often tagged as indices that 'race' exists as a legitimate explanation for disease prevalence differences among socially defined racial groups. Respondents in the study used the less pejorative term 'population' to imply that genetic compositional differences resulting from broad geographical factors correspond to 'races.'[17] Similar to 'race,' population as a classificatory concept approaches variable traits as a group rather than individual distinctions.[18] Rather than relieve genomics of 'race' as biology, the population concept maintains that racial differences are evident when genetic variants are entirely absent in one population and present in another.[19] This means that 'race' is squarely centered in the study of genomic variation through its reformulation as genetic populations. By substituting the term population for 'race' in genomics, population as a "less controversial and more technical-sounding genetic concept" affords respondents the means to evade considering racism impact on their research.[20] According to respondents, they often avoid use of the term 'race' in their research because:

> Because the term race has a long, a long and kind of nasty cultural history, right. I mean in the early twentieth century scientists who were talking about race were usually using it to like justify like horrible kind of eugenic programs, you know sort of horrible kind of behavior to their fellow man. Very few of my colleagues talk about race, per se. What they tend to talk about is population structure, and variation in allele frequencies around the world. That sort of tends to be the words they use.

Such comments reveal the discursive means through which respondents actively elude confronting the role of genomics in creating and sustaining racial reasoning and systemic racism. More importantly, interviewees' evasive racial discourse indicates their awareness of racialized reality rather than unconscious ignorance of it.

Despite this, interviewees like this individual persist in claiming, "I'm aware of having . . . to some extent racial prejudices . . . but I try not to act on them." But how can one not act, when racial ideology as a cultural constituent creates individuals and groups? As such, genomicists' views, perceptions, and interpretations of genetic variation are creations of their normative raced habitus. For instance, in order to define phenotypic variation, genomicists must specify a population. This *a priori* choice is necessary to construct a statistical model. Researchers' choice of population is often an important source of bias where hidden social assumptions enter genomic methods, given that genomicists use the 'race' concept as a cognitive map for constituting genetic populations.[21] Behind the population concept lies simplifying assumptions about how biology works and how genes and environment interact, all filtered through increasingly complicated and obtuse statistical models.[22] Relating genotypic and phenotypic variation depends on the populations chosen and the environments probed by genomics. The population concept as an abstraction is embedded with assumptions and prejudices. As mentioned earlier, when genetic clustering studies rely on samples taken from widely separated socially defined populations, they falsely suggest patterns of variation exist corresponding to racial categories. This occurs because the sample design of clustering studies often does not collect data from more evenly distributed geographical groupings. Sampling among members of geographically adjacent groups reveals genetic variation gradually changes from one group to the next.[23] This change in genotype and phenotype is not sharp but gradient instead, due to mating within geographically adjacent groups. Consequently, many individuals' genomes reveal a widely variable geographical range that links them to two or more 'continental groups.' Given this, the population concept merely serves as a terminological mask for 'race' in genomics. In the next section we explore how racial ideology is concealed through genetic racial euphemisms employed by respondents.

RACIALIZED CULTURE AS AN ENABLER FOR EUPHEMISTIC RACIAL GENETIC DISCOURSE

The intention to find versions of some genes of individuals in one 'race'—or 'ancestry,' a euphemism for 'race'—not found in members of other socially defined racial groups ignores the biosocial complexity of people, with all their unique character, history, and potential for change.[24] Although it is known with some degree of empirical certainty that human genome variation is an irreducibly complex phenomenon, many racially conditioned genomicists continue to assert that genetic frequency variation is the key to explaining phenotypical differences among socially defined racial groups. But the facts indicate otherwise given that phenotypes like skin color differences are most often products of genetic adaptation to eclectic environmental settings. As this scholar puts it:

> Dark skin evolved *pari passu* with the loss of body hair and was the original state for the genus Homo. Melanin pigmentation is adaptive and has been maintained by natural selection. Because of its evolutionary liability, skin color phenotype is useless as a unique marker of genetic identity.[25]

In more detailed terms, this neuroanthropologist explains, genetic change is

> linked to a system . . . of human cultural adaptability [thus not] manifested as distinct fine-tuned genetic adaptations in different races. . . . [K]ey human trait involves . . . adaptability to diverse contexts through learning. . . . Hence, an excellent explanation for variation across human populations is culture, not genes. Learning and enculturation explain most human behavioral variation. . . . Perhaps the best evidence is the simple fact that children adopted as infants across vast cultural boundaries do not carry their parents' social behavior with them.[26]

Notwithstanding this evidence, the notion that 'race' is a real biological fact continues within genomics largely due to the fact that the variation genomicists observe is not the basis of their interpretation, but racial strictures operating throughout genomics.[27] In racialized cultures human genetic variation is mentally organized in relation to phenotypic group dissimilarity. As such, genomicists conceptually make genetic meaning by contrasting and comparing socially defined racial groups' genes to one another. This happens even though genomic consensus maintains 'races' are complex sociocultural entities rather than genetically bounded categories. Informants exhibited this contradiction through their confusion about racialized language when discussing the population concept.

> I'm culturally from South Asia. I tend not to think of race. . . [We] use this concept of population based identification and Indian people identify themselves by language and the area of the country they come from and so, I'm an Indian who comes from Bengal. . . . I'm interested in population genetics. . . . And I've always understood that the kind of work [geneticists] do . . . has implications for the population at large.

These remarks illuminate how easily genomics as a racialized knowledge system can fill genetic terms like population with racial meanings. As such, racial logic and the idea of bio-'race' are sustained in genomics through newly developed genetic nomenclature like population, 'continental groups,' 'geographic origin,' 'ancestry,' and 'admixture' which implicitly stipulate racialized bodies are biological creations. These conceptions of 'race' as a thing with great alacrity encourage many researchers to falsely believe it exists in people's bodies not in their consciousness. This is why it is easier to alter the racial language for writing and speaking about human genome variation than to change the racial logic that genomics generates and sustains for investigating 'race' as a biological entity. The cultural permanence of racialism ensures that even when researchers forego the use of 'race,' their replacement terms for accounting for disease difference among human groups include racial mental associations.[28] 'Race' as a cultural sense-making classification serves not only to categorize/name group genetic differences, but also to influence the linguistic and mental basis through which genomicists create and maintain racialized cognitive frameworks for interpreting genetic information. As a result, genomicists' attempts to use less racially pejorative terms for explaining genetic variation among groups merely minimizes and silence racial ideology's centrality in new genetic variation language.

Similar to 'race,' the population concept is often used in human genetic variation studies without a clear definition. Many genome studies delineate a population's genetic similarity according to other ill-defined human group descriptors, such as sample donors' self-identified 'race' or ethnicity, nation of origin or regional identification, continent of ancestry, and population substructure.[29] The unspecified character of these descriptions is obvious in this respondent's explanation about how genomicists construct individual genotypes through comparison of socially defined racial groups' differences:

> . . . as a scientist we should focus on individual [not population] differences. [W]e have a large group of scientists now looking more into ancestry genomics, which is looking at how each individual may be different, you may have certain percent of African Americans, a certain percent of Caucasians, especially African Americans in America, . . . they're a mixture of populations. So they're doing a lot of whole, what we call whole genome studies (genome-wide association studies). And trying to make associations between those whole genomic associations and . . . certain diseases. Now you may get some data out of that that may be helpful, but when you look at that type of analysis, that's based on the, the profile and the genotype of those individuals. But when you convert that down to black, white, you won't be able to do that if you can say this person got 30 percent African American, 20 percent Caucasian, five percent Chinese . . .

Furthermore, the informant's words reveal how researchers conducting genome-wide association studies (GWAS),[30] who agree 'race' categories do not correspond to genetic variation, invoke notions of group genetic likeness to explain and evaluate genetic differences among groups.[31] This practice is unmistakable in researchers' use of population as stand-in for 'race' when answering questions pertaining to the importance of individual genotyping for the construction of *personalized* or *precision genomic* medical treatment:

> I hope that somewhere down the line in the future that we would be able to have a bank of drugs, each of which can be specifically targeted towards *particular population groups*. I think patients from *minority ethnic groups* would welcome wider representation of minorities across both biomedical research, clinical research, and therefore I think they would welcome a wider recognition on the basis of race, ethnicity [emphasis mine].[32]

The interviewee's equation of population to 'race' is not inadvertent given genetics' definition of population conceives of 'race' as a distinct genetic breeding group wherein particular traits are passed through generations within a socially defined racial group. In short, the population concept frames 'races' as variable traits of a group rather than individual genetic differences.[33]

Since the population concept is a central frame for making sense of genetic similarity, genomicists use 'ancestry' to account for genetic variation among socially defined racial individuals. Though 'ancestry' is a vague term with many different meanings and interpretations in different contexts, like population, it is calculated by inference of an individual's degree of genetic relatedness to an ancestral group.[34] Such reasoning is displayed in this researcher's words:

> But, if it's not recognized that there are differences in the, say, ancestry since the last 50,000 years of human history, it can lead to problems. So you could mistake an association between disease and genetic variance that could distinguish, say, continental African versus continental European ancestry or Asian. So, I formally divide my study subjects into self-described categories of people, based on whether they describe themselves as black or white. But I then do tests—genetic tests—that ignore [those self-described categories] and try to cluster what the ancestral backgrounds are of the different individuals. . . . [I try to do] formal tests to see if I can distinguish different groups of ancestry within a population.[35]

Given the ancestry concept's attention to how population history informs patterned genomic variation interpretation, respondents were quick to assert that their use of it is about genetic rather than racial genetic differences. Yet, their use of 'ancestry' ignores its sociocultural character as a concept dependent on *a priori* ideas about people's migration and geographic location for constructing genomic representations, interpretations, and meanings.[36] These sociocultural assumptions about group genetic likeness encouraged informants to assume individuals with similar genetic variation are members of a specific socially defined ancestral group. This logic is exhibited here:

> There are certainly genetic differences. And those genetic differences often have to do with ancestry. So there are many genetic variants that we can specify likely arose on one part of the world than another; and they are more common in the descendants of people who are born in one part of the world than another.
>
> So like we've known since the 1970s that if you pick any gene and find a genetic variance you can sort of ask . . . so for every sort of genetic difference you can ask sort of what is the probability that it's found in any old, random person versus what's the probability that it's likely due to what continent their ancestors came from? Or what's the probability it likely is due to where in the continent they're from? And the answer has always been about eighty percent of all variation is common to all humans. About ten percent of it looks like it has to do with what continent your ancestors came from. And about ten percent of it has to do with where on the continent your ancestors came from.
>
> So eighty percent of like all the variation in the world is just common to all humans. But ten percent of it is due principally to what continent you came from. And ten percent to where on that continent you came from. So because of that we can very easily distinguish often, which continent people have come from. We can sometimes even distinguish, you know, where on a continent they came from. It's often easy to tell someone who looks, you know, Greek from someone who looks say, Swedish. You know, there are sort of clear . . . kind of genetic

> differences. But that is due to a very, very small fraction of all the genetic variation in the world. We're just very good at recognizing it with our eyes.

As Joan Fujimura and Ramya Rajagoplan aptly observe, ancestry designations like the continental labels European and African tend to group together individuals with similar genetic variants, then attribute the similarities to shared descent from people from a single locale.[37] Since 'genetic ancestry' inferences are dependent on reference population data for comparative purposes, many informants held the view that people with similar genetics are members of a specific labeled population thus share related ancestral geographic origins (Africans, Asian, European, etc.) or sociocultural labels. Given this conceptualization, interviewees frequently used socially constructed descriptors like 'national labels' and 'regions of continent' to evaluate genome variation similarity as evident in this exchange:

> Interviewer: Do you think that focusing on genetic variations among human groups is important?
> Interviewee: Yes, of course, it's very important.
> Interviewer: Okay. Explain why.
> Interviewee: It's because there are lots of race specific knowledge which we are harnessing. So, for example, there is a project going on funded by NIH and other funding agencies called International HapMap Project. The purpose there is to seek thousands of individual DNA sequence of different ethnicity. So it's imputing the ones which we never studied for example, Indians, Southeast Asians, [inaudible] Chinese population and several different subgroups of tribes from Africa and there are even among Caucasians there are two or three different subgroups which we have included. There are generic variations between these subgroups and then you can correlate this with some, let's say, one group of ethnicity has more predisposition to a particular disease.

Though respondents acknowledged that genetic variance that is differently distributed across socially defined 'continental'/'ancestral' groups can falsely intimate biological 'races' exist, most asserted, like this researcher, that this occurs

> Very, very rarely if at all. . . . [F]or the most part we . . . So for the most part we try . . . [to] look at almost the entire genome when we're doing studies. So when we . . . deal with these differences in the allele frequencies, we look directly at the alleles and look at the differences in the allele frequencies between humans. And often we do it explicitly without asking [people any questions] about where they think their ancestors come from, what race they would describe themselves. We often explicitly *ignore* that information and just look exclusively at what the genes are telling us.

Despite informants' assertions about disregarding 'race,' their inferences of shared ancestry on the basis of genetic variation similarity via the continental categorization of groups ensures their genome interpretations and understanding are racially inflected. The implication of racial thinking in statistical probability is apparent among genomic statisticians whose inferences are integral to how they make sense of genome variation among socially defined racial groups. As this statistician put it:

> [Race] is central to what we do. . . . We try to handle it in different statistically sophisticated ways, but there's self-report genetic evidence, and it all gets . . . mixed up. We're doing a big genome-wide association study on metabolic factors in cardiovascular disease; metabolomics. Despite the fact that we haven't observed big racial differences in some these metabolites, we're still stratifying on race and then doing a meta-analysis. I would like to explore ways of doing it better, although there is definitely . . . a way to do it in the GWAS context . . .

Most respondents refused to accept that continentally demarcating groups operates as a euphemism for racial identifiers in genes. Rather than clarify how different ecological locations induce a vast majority of human genetic variation in disease susceptibility difference, informants reverted to racial-biological reasoning to explain the differing disease risk among groups. The importance of environmental factors in disease risk is unmistakable in Cooper and collaborators' research conclusion that people of African descent are not intrinsically susceptible to high blood pressure.[38] Specifically, Cooper and colleagues found that the inherent racial genetic reductionist character of genomics encourages researchers to oversimplify complex human genome variation processes like hypertension to stress the importance of endemic biological racial factors rather than the effects of racism and other sociopolitical/environmental variables in increasing a group's susceptibility to disease.[39] The researchers reasoned how else could this explain why 'white' Euro-Germans as opposed to Nigerians possessed higher rates of hypertension. All of this suggests the consistent emphasis given to the genetic elements of racial contrast diverts attention away from the more important problem of defining and intervening to prevent disease that are similar in impact regardless of an individual's ethnic and racial background.[40]

Because most interviewees were unreflective about racial thinking in genomics, they tended to uncritically accept the use of 'race' as a genomic variation descriptor. This respondent's words typify this taken-for-granted practice:

> As scientists, we can . . . clarify that race is a complicated term that includes genetic components such as color and physical features (shapes of noses, mouths, and eyes, for example), non-genetic components such as nationality, language, clothes, and customs, and components that are both genetic and environmental components, such as

> diabetes in Pima Indians. It is clear that Pima Indians have a genetic predisposition to develop diabetes, but only developed the disease in response to a change in their diet. Rather than denying genetic differences, we as scientists can guide conversations about race so that they first presume tolerance, and then are more scientifically precise.[41]

Notice how the respondent acknowledges but then proceeds to ignore the significance of the gene-environment dynamics for the Pima's diabetic susceptibility by emphasizing the overriding importance of a preexisting genetic racial disposition for diabetes. Racial biological reductionism of this sort, as mentioned beforehand, is largely attributable to population assumptions suggesting that individuals are more likely to be susceptible to a disease if many members of their socially defined racial groups carry a similar determinative genetic trait. Such reasoning runs counter to the fact that on average humans are more genetically similar than different and no two are biologically identical.

As Bourdieu's habitus concept suggests, genome researchers' words are politicized even when they not aware of it because their words and interpretation of biological reality reflect the interests of their immediate social and historical environment.[42] Accordingly, no matter what new genetic variation categories genomicists develop to avoid acknowledging racial ideology's influence in genomics, they and their work's immersion in racialized sociohistorical contexts incline them to draw heavily on existing racial meanings to interpret genetic data.[43] For this reason informants were less likely to consider how 'race' and racial genetic euphemisms like 'geography' racially simplify and distort their interpretation of genetic variation. This is exemplified in this respondent's words:

> I'm not even sure what race means [because] people use it in many different ways. In our own studies, to avoid coming up with our own definition of race, we tend to use the definition others have employed, for example, the U.S. census definition of race. There is also the concept of the major geographical structuring that exists in human populations—continental divisions which have led to genetic differentiation. But if you expect absolute precision in any of these definitions, you can undermine any definitional system. Any category you come up with is going to be imperfect, but that doesn't preclude you from using it or the fact it has utility.[44]

Despite the fact that 'race' as a genetic proxy in genome research seldom reveals fundamental new understandings of disease, respondents' socialized racial ways of knowing undercuts their ability to see human genome variation as normative uniformity. To truly grasp the complexity of human genetic homogeneity, genomicists need to refocus their efforts on determining how multitudes of genotypes, that are basically the same, contribute to the manifestation of similar diseases. This is starting to be done through the new genetic subfield of epigenetics, which examines how external environmental factors switch genes on or off in ways that determine whether genes are expressed as disease. Although this is the direct opposite of inferring genetic risk for disease through racial taxonomy, epigenetics still leaves the door open for racialization through its contention that epigenome changes can be inherited in humans. Accepting that you cannot put people neatly in a small number of meaningful racial and ethnic categories like 'black, white, African, European or Asian' infers genetic variation is not easy for most genomicists situated in a racialized culture as this informant voices:

> . . . after reading all the [debates] in the literature—I'm like—the people who say that you shouldn't study race, and all that, I'm against that. Maybe studying race can give you a lot of insight. So all these complex disorders [like asthma], why not compare different races with different prevalence of diseases and comparing them might give you some clue that they are different—at like, at say, at these chromosomal potions—why not study them, right, just compare? Do this practical thing, just compare and that can give you some clue that can narrow down your search for *genes for asthma* (emphasis mine). . . . I think the race thing people use is [for] people to speak the same language and [have] the same social background. . . But, *then*—then the thing is when you divide Mexicans and Puerto Ricans—Mexicans and Puerto Ricans have the same—they speak the same language and they have almost the same social set-up, so they're the same, so—but then Mexicans and Puerto Ricans have the very different, say asthma, they're very different . . .[45]

Respondents used ethnic descriptors and terms like 'Caucasian' and 'population' in a manner that implied they accepted social suppositions about the correlation of 'race' to genetic variance. Language like 'Asian genes,' 'African mutations,' 'African or European alleles' or 'ancestral allele' signified thinking among interviewees that intimates that distinct types of genetic racial differences exist. Interviewees, like this individual, used these terms in conversation to construe small amounts of genetic divergence implied natural distinction among socially defined racial groups exist:

> Only one variation in SLC24A5 changes amino acid. To our delight, we noted that that variation is associated with extreme frequency difference between African and European populations. In the European HapMap population the ancestral allele has a frequency of zero, whereas both the African and two East Asian populations showed an ancestral allele frequency of 97 percent–98 percent.[46]

Admixture is another concept that interviewees used to imply preceding genetic uniformity among socially defined racial groups.[47] Genetic admixture links the proportion of African, European, and Native American ancestral genetic makeup of individuals to increased disease susceptibility. Accordingly, the genetic admixture approach makes several assumptions. First, genetic variation in groups differ markedly from one another primarily due to geographical barriers; second, when individuals of genetically distinct groups engage in significant intergroup mating, differences are diminished; and third, the 'percent admixture' is medically important if individuals' inherited genes are known to play a role in particular medical conditions and if those genes are known to differ in socially defined racial groups.[48] As such, the genetic admixture concept presumes genetic 'purity' is healthy while 'ancestral' genetic mixing increases a group's disease susceptibility—the very assumptions eugenics used to legitimate implementing racist reproduction controls. Thus admixture assumes that innate ancestral biological determinants explain socially defined racial group differences in susceptibility to disease. Although respondents using the genetic admixture approach contend it considers both the nongenetic and genetic contribution of racial social groupings, it was clear their focus was on neither environment nor the social determinants of disease etiology. Rather than consider how varying differences in environmental exposure contribute to increased disease susceptibility, respondents' admixture discourse led them to recreate biological 'race.' This was done through the naturalization of disease differences on the basis that genetic distinction explains the differences in disease pathologies among socially defined racial groups. Respondents thinking about disease risk in this way articulated it so:

> I do know Mexico is trying to get a better understanding of how race and racial admixture influence disease expression. My group has proposed a study of this topic to the Mexican government and they have expressed some interest in pursuing it. This is because Latinos are a mixed population of Europeans, Native peoples and Africans. This study will help us gather important information about how disease is expressed in populations with diverse ancestries. My group has published a number of papers showing that disease risk varies depending on a person's racial background. For example, the more Native American you are, the milder the asthma. However, the more European you are, the more severe.[49]

Such genetic admixture reasoning sorts people into social categories presumed to be 'races' that are not only based on skin color or other visible attributes, but also on genetic predisposition to disease.[50] Given this, more often than not, descriptions for risk factors for disease often include statements using racial group descriptors as proxies for disease susceptibility. For example, the U.S. Prevention Service's statement on prostate cancer screening risk factors reads:

> Older age is the strongest risk factor for the development of prostate cancer. However, neither screening nor treatment trials show benefit in men older than 70 years. Across age ranges, black men and men with a

> family history of prostate cancer have an increased risk of developing and dying of prostate cancer. Black men are approximately twice as likely to die of prostate cancer as other men in the United States, and the reason for this disparity is unknown.[51]

When interpreting genome differences, informants racialize variation via genetic markers identified as being linked to an increased risk of disease in certain socially defined racial groups. This association is established initially through phenotypes/appearances, which presume genetic homogeneity in phenotypically similar groups thus ignoring the complex web of biological and social connection linking individuals and groups to each other and illness. In light of this, though respondents were aware of how to think systemically about human genome variation, the systemic thinking they employed usually interpreted genotype differences as denoting different kinds of human groups.

GENETIC RACIALIZATION OF DISEASE

Racialization in genomics occurs when distinctive genetic signatures are imagined as existing within groups labeled as members of a specific population. The primacy of genetics as a determinative factor helps to sustain the fallacious idea that 'race' is biological and a crucial reference for understanding disease susceptibility differences among socially defined racial groups. This logic is implicitly at work in this Alzheimer's Association write-up about African Americans' higher risk for the disease:

> A newly published study identified a gene that may increase the risk of developing Alzheimer's disease among African-Americans. Until this study, no known genetic factors could account for the greater risk. And even now, the study's findings must be replicated before the conclusion can be validated, and the genetic study did not address the higher prevalence rate among Hispanics. Even if there is a genetic link, other factors are also likely to play a role. Conditions such as high blood pressure and diabetes, both of which are risk factors for Alzheimer's and other dementias, are more common in African-Americans and Hispanics than in whites—and probably account for some of the differences in prevalence rates. Similarly, lower levels of education and other socioeconomic characteristics (such as lower income) among older racial and ethnic minorities may also contribute to increased risk. In fact, *some* studies have shown that differences in prevalence by race and ethnicity largely disappear when these factors are taken into account.[52]

Though this disease narrative attempts to downplay the centrality of the racial genetic link by incorporating nongenetic factors for susceptibility, inclusion of racialized illnesses like high blood pressure and diabetes encourage reading Alzheimer's as a 'race'-specific disease. The inscription of 'race' onto genetic markers associated with diseases like Alzheimer's occurs in spite of informants' awareness of this possibility as demonstrated in this respondent's carefully worded explanation:

> I think what's important is tracking down the genetic basis of human disease, and complex human traits, but I don't think trying to break those up by some kind of ethnicity or racial boundaries is useful at this point, or helpful. Because I think that you know, there's a real danger of introducing a bias, and given the similarity of the genetic makeup of humans from around the world. . . . I think the first thing that's important is, can we identify genes that are responsible for complex disease? Can we identify genetic constitutions that contribute to mental illness? Those are the issues to go after . . . You know where you can really see an outcome for possibly . . . developing treatments or therapies based on that information. But I don't see a reason to . . . it doesn't seem like an important thing to me . . . given [our genetic] commonality, to be . . . look[ing] for small differences between different ethnicities and races.

These comments intimate there is a strong belief in genetics as the primary source for disease and a need to explain group disease susceptibility differences as a manifestation of the genetic sameness of members of socially define racial groups.

Critical to the search for genetic markers for disease is the population chosen to study for appearance of the illness. It is here, as mentioned earlier, that some prior knowledge of a group's disease history becomes a part of genetic association studies testing for a correlation between group disease status and genetic variation.[53] These associations produce racial genetics through the identification of gene frequency variation in socially defined racial groups linked to disease susceptibility. This racialization process is evident in a recent study referred to in the Alzheimer's Association's risk statement in a GWAS analysis of Alzheimer's disease in 5,896 African Americans over age 60.[54] About 2,000 participants actually had Alzheimer's disease, and the rest who were included for comparison did not. The study identified the same gene variants in older African Americans found in older people of European 'ancestry.' But they found that African Americans with Alzheimer's disease were slightly more likely to have one gene, ABCA7, thought to confer risk for the disease.[55] The study's researchers calculated that ABCA7 increased Alzheimer's risk by about 80 percent in African Americans, compared with about 10 percent to 20 percent in people of European 'ancestry.'[56] In this way, interviewees' genome data collection from specific racialized populations encouraged them to look for genomic sequence variation in these groups in manner that quantified and sorted the Alzheimer's genotype into biological 'races.'[57]

Such racial sorting occurs despite genomicists knowing that the difference in risk between African Americans and those of European descent is likely tenuous. They know that genetic links, like that with ABCA7, are often more significant in some studies and less in others simply because populations vary purely by chance and environment variability; yet researchers still seek to identify how disease susceptibility variants segregate into socially defined populations.[58] Moreover, while the object of their analysis is genetic, genomicists use racial labels to describe the groups being evaluated to guard against confusing a chance difference in gene variance with disease etiology.[59] That is, genomicists do not generally situate genetic susceptibility with membership in a given socially defined racial group; rather they summarize the associations as observed across different populations. However, interviewees often assumed that human morphological features are correlated with deeper traits and racial labels are crude genetic descriptors of human groups. As such, this respondent presumes

> that you should treat . . . all the patients medically the same way regardless of race, because they're all people and they should all react the same way. [But] the fact is that different racial groups do have different genetic make ups and different genetic backgrounds. There's some diseases that are more common among people of African ancestry and some diseases that are more common among people of Caucasian ancestry. To ignore that as a geneticist or a medical professional is silly in the same way that you, that it would be silly to ignore, you know, smoking history if you're (inaudible) lung cancer. But at the same time, to say th[is] among non-geneticists makes it sound as though you're saying that the color of somebody's skin affects their disease status, not their genetic background.

Notwithstanding the informant's contention that genomicists do not use socially defined 'race' to infer an individual genetic disease status, the 'race'-gene correlation is implied in genomic methodology and language suggesting differences in disease susceptibility risk linked to genetic variances among socially defined racial groups.

Though respondents are well aware genetic racial contrast is imprecise, their searched for preexisting genetic susceptibility differences among socially defined racial groups is centered in racialized notions about these group's genetic predispositions. Their efforts to genetically differentiate socially defined racial groups must not be understood as merely a characteristic of individual genomicists, but a product of an all-pervading racist cultural milieu that encompasses their genomic thought and action. This racialized sociocultural system operates as a nebulous haze inhibiting genomicists' capacity to approach disease as biological embodiments of sociopolitical processes and lived experiences.

NOTES

1. My approach is similar to Obasogie's research evaluating people's racial sensibilities reproduce and sustain racism in his book *Blinded by Sight: Seeing Race through the Eyes of the Blind*.

2. Miles, *Racism.*
3. Miles, *Racism.*
4. Informants often used Caucasian either as a self-identifier or a label to designate the genome of others they categorized as such. The term is problematic because it postulates that the natives of the Caucasus exhibit idealized physical appearance so the Caucasus are believed to be the birthplace of humankind. The logic behind this idea—the assumption that 'whites' exhibit the best physical appearance—is implicitly racist. Additionally, we now know our species first appeared in Africa, so the biology is not sound either.
5. Lipsitz, "The Possessive Investment in Whiteness: Racialized Social Democracy and the 'White' Problem in American Studies."
6. Woods, "5 Things No One Is Actually Saying About Ani DiFranco or Plantations." *Scott Woods Makes Lists blog*, January 3, 2014, http://wp.me/p40Hhl-1g.
7. Foucault, *Power/Knowledge: Selected Interviews and Other Writings.*
8. Leroi, "A Family Tree in Every Gene."
9. Rosenberg, et al., "Genetic structure of human populations," 2381–2385; Risch et al., "Categorization of Humans in Biomedical Research: Genes, Race and Disease."
10. Thompson, "The Problem of 'Race as a Social Construction.'"
11. This technique for racial categorizing is also being used by sociologists trying to link 'races' to biological groups. See: Morning, "And You Thought We Had Moved Beyond All That: Biological Race Returns to the Social Sciences."
12. An allele is a gene (one member of a pair) that is located at a specific position on a specific chromosome. Alleles may occur in pairs, or there may be multiple alleles affecting the expression (phenotype) of a particular trait.
13. Morning, "And You Thought We Had Moved Beyond All That: Biological Race Returns to the Social Sciences."
14. Jorde et al., "The Distribution of Human Genetic Diversity: A Comparison of Mitochondrial, Autosomal, and Y-chromosome Data."
15. Dunham, "Genetic Links to Prostate Cancer."
16. Fausto-Sterling, "Refashioning Race: DNA and the Politics of Health Care."
17. Wald, "Blood and Stories: How Genomics is Rewriting Race, Medicine and Human History."
18. Gannett, "The Biological Reification of Race."
19. Gannett, "Racism and Human Genome Diversity Research: The Ethical Limits of 'Population Thinking.'"
20. Gannett, "Racism and Human Genome Diversity Research: The Ethical Limits of 'Population Thinking.'"
21. Pankaj Mehta, "There's a Gene for That."
22. Pankaj Mehta, "There's a Gene for That."
23. Serre and Pääbo, "Evidence of Gradients of Human Genetic Diversity Within and Among Continents."
24. Pollack, "Natural Selection, the Human Genome, and the Idea of Race."
25. Jablonski, "The Evolution of Skin and Skin Color."
26. Lende, "Nicholas Wade and His Determinist Genes."
27. Saussure, *Course in General Linguistics.*
28. Fujimura and Rajagoplan, "Different Differences: The Use of 'Genetic Ancestry' versus Race in Biomedical Human Genetic Research."
29. Fujimura and Rajagoplan, "Different Differences: The Use of 'Genetic Ancestry' versus Race in Biomedical Human Genetic Research."
30. Genome-wide association studies search the genome for small variations, called single nucleotide polymorphisms or SNPs (pronounced "snips"), that are thought to occur more frequently in people with a particular disease than in people without the disease. Each study can look at hundreds or thousands of SNPs at the same time. Genomicists use the data to pinpoint genes that may contribute to a person's risk of developing a certain disease.

31. Fujimura and Rajagoplan, "Different Differences: The Use of 'Genetic Ancestry' versus Race in Biomedical Human Genetic Research."
32. Malik, "Analysis: Colour Coded Prescription."
33. Gannett, "The Biological Reification of Race."
34. Fujimura and Rajagoplan, "Different Differences: The Use of 'Genetic Ancestry' versus Race in Biomedical Human Genetic Research."
35. Fujimura and Rajagoplan, "Different Differences: The Use of 'Genetic Ancestry' versus Race in Biomedical Human Genetic Research."
36. Fujimura and Rajagoplan, "Different Differences: The Use of 'Genetic Ancestry' versus Race in Biomedical Human Genetic Research."
37. Fujimura and Rajagoplan, "Different Differences: The Use of 'Genetic Ancestry' versus Race in Biomedical Human Genetic Research."
38. Cooper et al., "An International Comparative Study of Blood Pressure in Populations of European vs. African Descent."
39. Cooper et al., "The Puzzle of Hypertension in African-Americans."
40. Cooper et al., "An International Comparative Study of Blood Pressure in Populations of European vs. African Descent."
41. Glaser, "Investigator Profile: An Interview with Keith Cheng, M.D., Ph.D."
42. Bourdieu, *Outline of Theory and Practice.*
43. Hunt and Megyesi, "The Ambiguous Meanings of the Racial/Ethnic Categories Routinely used in Human Genome Research."
44. Gitschier, "The Whole Side of It – An Interview with Neil Risch."
45. Fullwiley, "Race and Genetics: Attempts to Define the Relationship."
46. Glaser, "Investigator Profile: An Interview with Keith Cheng, M.D., Ph.D."
47. Jackson, "Anthropological measurement: The Mismeasure of African Americans."
48. Fausto-Sterling, "Refashioning Race: DNA and the Politics of Health Care."
49. Johmar, "Interview Subject: Dr. Esteban González Burchard."
50. Fee, "Racializing Narratives: Obesity, Diabetes and the 'Aboriginal' Thrifty Genotype."
51. U.S. Preventive Service Task Force on Prostate Screening, "Final Recommendation Statement."
52. Alzheimer's Association, "Race, Ethnicity and Alzheimer Disease."
53. Al-Chalabi and Almasy, eds., *Genetics of Complex Human Diseases.*
54. The study citation: Retiz et al., "Variants in the ATP-Binding Cassette Transporter (ABCA7), Apolipoprotein Eε4, and the Risk of the Late-Onset Alzheimer Disease in African Americans."
55. Kolata, "In Blacks, Alzheimer's Study Finds Same Variant Genes as in Whites."
56. Kolata, "In Blacks, Alzheimer's Study Finds Same Variant Genes as in Whites."
57. Fullwiley, "The Biological Construction of Race: 'Admixture' Technology and the New Genetic Medicine."
58. Kolata, "In Blacks, Alzheimer's Study Finds Same Variant Genes as in Whites."
59. Paradies et al., "Racialized Genetics and the Study of Complex Diseases."

FOUR

Racialization via Assertions of Objectivity and Heuristic Practice

Given that genomicists' observations are a part of their racialized experiences in the world it is virtually impossible for them to collect and interpret biological facts without intervening racial bias. In view of this, genome variation explanations are underdetermined by evidence in that genomicists' particularized assumptions mediate the relationship between gene activity and accounts of this biological process. There are numerous cases throughout genetic history, as I discussed in chapter 1, where scientists' background beliefs distorted their genetic observations. These biased readings arise, not merely out of deception, but from genomicists' deep submersion in commonsense racial ways of knowing, beliefs, and attitudes instructing their thinking about human genome variation. Correspondingly, these commonsense racial ways of making sense of human differences ensure that certain genome variation facts are either not considered at all or are deemed unimportant. Since these ways of knowing are internalized and shared with the broader society, they are not simply personal but external social facts structuring researchers' assessments of human genome variation. Racial beliefs' ordinariness and uniformity means that more often than not racial reasoning in genomics is rendered less visible thus unavailable for critical scrutiny. In what follows, I discuss how 'race' is made natural and invisible through genomicists' 'objectivity' claims and heuristic practices.

OBJECTIVITY AS A RACIAL TROPE

In trying to clarify how racial ideology is insinuated in genome science, we must begin by taking stock of the particular epistemic values genomicists bring to their research. The basic foundation of science is that it deals with facts, not values, thus scientific investigations are value-free or 'objective.' This principle offers scientists the ease to believe their socially constructed methodologies provide them with a bias-free way to describe reality. But, scientific 'objectivity' is a particular and specific sociohistorical conception that does not transcend social influences.[1] Contrary to the traditional way of understanding 'objectivity' in science as a kind of transcendent "conquering gaze from nowhere," that positions scientists as distant from and outside of the object of inquiry, 'objectivity' is an illusory cultural construct that conceals the ideological basis of a supposedly disembodied scientific objectivity.[2] In Karl Mannheim's words: all knowledge systems "are not taken at face value but are interpreted in light of the life-situation of the one who expresses them" because all-knowing subjects are reliant on explicit socially conditioned outlooks to make sense of reality.[3] This also means that people's experiences in the world, as standpoint theory asserts, differ in social location and time in ways that inform how they perceive and interpret objective reality. This conception of knowledge production, accordingly, suggests that 'white' genomicists, who predominant in the discipline, stand in different epistemic relations to other genomic inquirers—for example, African Americans and Latinos—given that their ideas about the human genome are a function of their position in the social milieu. As evolutionary biologist Stephen Jay Gould remarked, this all means "objectivity cannot be equated with mental blankness; rather, objectivity resides in recognizing your preferences and then subjecting them to especially harsh scrutiny—and also in a willingness to revise or abandon your theories when the tests fail (as they usually do)."[4] As a socially embodied cultural creation, 'objectivity' offers genomicists a powerful way to imagine 'race' in the human genome in a manner that leaves the unmarked 'white' power to see and not be seen, to represent while escaping representation in genomics.[5] Whiteness is not treated as a 'race,' and thus is made invisible and not a concern in genomics. 'Objectivity' contentions regarding human genome variation research then are about racially situated genomicists, not a value-free analysis of the genome. Genomicists' 'objectivity' claims obscure how their sense making signifies their perverse capacity—cultivated within the contexts of systemic racism—to distance themselves from everybody and everything in genomics that sustains racial oppression.[6]

Interviewees found it difficult to acknowledge their racial position's influence on their genomic interpretation given their belief in the capacity of the scientific method—systematic observation, critical reasoning, and extensive peer reviewing—in minimizing the effects of social, personal, and political elements in their research. This reluctance was detectable in this informant's words:

> [I]n my reckoning, people don't let those biases cloud their analysis of material and I'm not talking about sociologists and the philosophers and things like that. I'm talking about hard core human biologists and scientists and I'm talking about at least the physicians I've come in contact with. And I think one thing scientists strain to do is to be relatively objective about the information they have. Now their views are clouded by the paradigm constraints of whatever period of history you're embedded in. There's no question about that. But I don't think most scientists let racial perceptions cloud their ability to analyze data.

Similarly, this respondent pushes back the suggestion that genome science is value laden by emphasizing the importance of mathematic probability analyses in securing a bias free genomics:

> [M]ost of the analysis we do is actually statistical. And I'm not sure. I mean I guess you could say statistics is a language, but I mean most of the way we analyze the data are in terms of statistics. And, so I don't think [social influences are significant]. You know, certainly not consciously. We actually, I think actively try to avoid preconceptions [seeping into] data. So, I think not, you know, from just thinking about it . . . in the time since you've asked the question. You know, that would be my initial thought because we do try to eliminate those kind of things from the way we . . . we try to minimize their presence in the studies and then try to also minimize their presence in the analysis.

Though it is highly implausible for the scientific method to be 'objective' given its mediation by racialized cultural prescriptions, respondents expressed an abiding faith in science's ability to produce some degree of value neutrality. A sentiment one informant verbalized this way:

> I don't think it's a black way of thinking about science or white way. I don't think there an American way of doing science compared with a Chinese way. Science is a logical construct and the way one goes about doing this is either effective or not effective. I would be very uncomfortable agreeing to the notion that there's a white way of doing science or thinking about science and the black way, European versus American or so forth. I don't think there is a cultural influence. To the extent there is culture that is influencing my scientific career, it might have influenced my going into a scientific career to begin with. That is, I grew in a family where intellect and learning was highly respected and that clearly an effect in making it [easier], making it more conducive for me to go into a career—and academic career involving science. But once I got into the academic career of doing science . . . my background and my culture [did not] ha[ve] much influence on the way I d[o] science.

Some interviewees voiced the confused view that new statistical models in genomics are both free of racial bias and involved in them. Such muddled thought reflects the fact that new scientific methodological technologies are embedded with racial signification through comparison of populations that translate seemingly 'objective'/neutral digital statistical algorithms into 'bio-races' that was not obvious to informants as is apparent in this exchange:

> Interviewer: What are the social implications of the research you're doing in Computational Genetics?
> Interviewee: So probably, formally, for me not much. Because it's sort of, being almost technical about it, what I mostly do is I often build the statistical and computer models used to analyze data. But I don't often analyze data, per se, myself. Or make conclusions from it.
> Interviewer: Okay.
> Interviewee: So I often build a tool that someone else might be able to use to make some kind of cultural inference from. But I really don't.
> Interviewer: All right, but you are very much aware of how genetics works, right?
> Interviewee: Yeah, very much so. But personally, it doesn't come up in my work on a day-to-day basis.
> Interviewer: So do you think . . . when you're putting together this kind of like sequencing technique, you are not influenced in any way by racial conceptions of reality?
> Interviewee: So . . . what we're usually interested in doing, is finding . . . variance in genes that are associated with disease. And in our work, the . . . key thing to distinguish is . . . find[ing] variances that differ . . . between people who have disease and don't have disease. But not because they differ because [a person's] . . . ancestors [are] from one part of the world rather than another part of the world. So the notion [that] genetic differences between people whose ancestors [a]re from different parts of the world . . . [is] like a nuisance complication that prevents me from making, often, a very easy determination whether or not a gene is associated with disease.
> Interviewer: Okay.
> Interviewee: And we're sort of constantly trying to deal with figuring out whether the allele differs because people have different ancestors or whether the allele differs because it's actually associated with disease.

Though many respondents did acknowledge their work is susceptible to societal influences, they carried on their inquiry as if history and racial ideology are not intrinsic significant factors in their research. This is demonstrated in this interviewee's response to a question on whether not racial ideology impacted his understanding of genetic data:

> Interviewee: Not really, no. I think most of my professional life was not in India. So whereby I mean I finished part of my PhD in London and finished all my other education and clinical fellowship and everything in the US. I [attended] Harvard Medical School. So I think, if at all, it makes me a little bit more conservative and really think harder about the implication of my reporting in the diagnostic field. But I don't see that it's any different than any other ethnic or nationality.

> Interviewer: I guess what I'm trying to get at is, do you thing that any of the kinds of cultural and social cues that you were brought up with, do you think that they in some way unbeknownst, even consciously intercede in how it is you analyze the data, the genetic variation data?
> Interviewee: That's a good point. But what I'm trying to say is that, the only thing that I can think of is that I'm a more, maybe more conservative in looking at, double checking all the data and trying to find more evidence. But I don't think that, I'm in no position to say that my colleagues who are not of the same upbringing are not doing that. You know what I mean?
> Interviewer: So what you're saying to me, or at least what I am hearing is that you are constantly concerned about data validity, therefore you make sure that you cross check the data over and over again and make sure that's what you observed is reliable and correct.
> Interviewee: Yes. But having said that, I'm not sure if the person next to me doing the same kind of job and not doing it, you know?

Similar reasoning is expressed outside the United States in the United Kingdom where the House of Lords 2000 Science and Society report asserted that "though science in itself is neutral, the application of science is not."[7] Belief in science's neutrality led many respondents not to spend much time thinking about sociopolitical influences in their work, thus avoiding consideration of these factors in a way that maintained the pretense of 'objectivity.'[8] For as this informant emphatically put: "I'm very much a scientist that believes in fact not opinion. The interpretation is very simple. I know which mutation caused the disease, and the severity of the mutation determines the severity of the disease and how much treatment to give." Informants expressing this belief were usually strong proponents of approaching disease as an outcome of genetic difference. One astute respondent maintained that genomicists' preference for genetic explanations is a consequence of the fact that "geneticists have strong training in genetics and very weak training in the environment[al]" etiologies of disease. This predilection is also driven in part by informants' conviction that though they themselves are not shielded from societal influences, genome science is *pure*. An outlook embodied in this respondent's comments about racial discourse in genomic data interpretation:

> Well you know you sometimes get into some of these conversations. But mostly we talk science. You know science to science. But it will come up, but mainly due to the lack of people not studying the field. You know we're trying to get more young people in the field, from that aspect. So that interpretation, you know, you want to train people the correct way.

This conviction that scientific methods are beyond sociopolitical reproach was widely shared among respondents who like this interviewee stated:

> Am I concerned? Scientists always ask themselves, "What are my assumptions?" We're taught to try to continually examine those assump-

> tions, and whether or not the data bear them out. . . . Just to be able to say . . . when I write a scientific paper and I have to describe my donors, what is the word that I use.
>
> If I'm trying to accurately portray what the history of a population is, of course I'm concerned [about culture's effect on my assumptions]. Sometimes we see unusual results, then the question becomes, were those assumptions correct? Especially when you start asking about humans and family history, because we lie.

Respondents' assumption that genomic methodologies are value neutral enabled them to ignore how genomic variation research and racism co-produced one another. Such evasion was routine given that acknowledging white supremacy's influences on their work required them to concede:

> [objective reality] acquires meaning only through human definition and engagement that include a variety of possible mediations and individual perceptions. Ideas, concepts, the theoretical enterprise itself, are all part of a historically-evolving socio-political process and, as such, resist simple determinist models. Methods and techniques of analysis can serve little function independent of their historical content and purpose.[9]

Thus rather than admit that scientific objectivity is an illusion, interviewees like this one used the pretense of 'objective' reasoning to contend:

> Our society, given its sordid history on race related issues, is very confused about how to deal with racially and ethnically sensitive topics. *As a result, science and politics get mixed up when they relate to these topics. I personally feel, like many other scientists, that science should be separate from politics* [emphasis mine]. In particular, science should meet the same burden of proof regardless of what political implications it might have. But this may be too idealistic if not naïve.[10]

In a succeeding interview the respondent candidly stated that some scientists "start with a political agenda and fit the evidence to that." This political bias, he continued, "takes credibility away from an antiracist program I agree with. . . . If someday we discover that there are genetic differences in cognitive abilities, would that mean that racism is now justified?"[11] Rather than critically challenging the veracity of the 'race' concept in genomics, respondents' assumed their use of 'race' would help alleviate practical problems such as health inequalities. 'Objectivity' discourse also enabled interviewees to characterize genomics as apolitical 'science,' thus there was no need to account for how racial ideology prefigures genomic knowledge.

Because structurally situated genome researchers are very likely to confirm or see whatever their racialized social structures designate as reality in their work, they are disposed to depict disease susceptibility differences among human groups in ways that scientifically instantiates variation as bioracial distinctions between them. Respondents produced the 'scientific truth' of racial biological differences through, as we have seen, research designs grounded in racial typological logic embedded in statistical reasoning, where variance between human groups is relative or clinal. That is, gene frequencies gradually change over a geographical area usually as a result of environmental fluctuation. Though there is no natural divide separating the groups into 'races,' the genetic difference between geographically distant groups are assumed to differ to the point where they could be interpreted as biologically distinct. In this way, researchers interpret genome evolutionary/adaptive processes to demonstrate 'race' is a biological concept. Given this, it is no longer necessary to say there are racial typological distinctions between human groups because gene frequencies differences and probability authenticate 'race' as biology.

Inferring bio-'race' statistically derives from the mathematical frame eugenicists Francis Galton and Karl Pearson bequeathed to genetics. They use their supposedly value-free statistical algorithms, as I delineated in chapter 1, to scientifically justify ranking ordering people in racial and class groups. As Nadia Abu El-Haj's recent writings reveal, genomic statistical analyses project neutrality through their use of probability calculation and risk assessments, avoiding the label biological reductionism by organizing around the notion of the "potentiality for illness."[12] In other words, susceptibility for disease is not inferred so much from tagging some person or groups as pathological because numerous individuals exhibit a disease, but from the fact their bodies possess a genetic mutation that could potentially develop an illness. Though risk logic deviates from genetic determinism by contending the presence of variance does not always translate into disease expression, the genetic profile of persons or groups identified as possessing specific genetic markers for disease can be racially construed to insinuate socially defined racial groups identified as carriers are inherently 'unhealthy,' thus racially distinct.[13] This occurs primarily because statistical probability calculation about the correlation between genotype and phenotype is a social creation interpreted through racialized human cognition. Fullwiley, for example, found nineteenth century conceptions of racial groupings are used in genomics to create statistical thresholds for a series of genetic markers to constitute populations (read: 'race').[14] These statistical models compare gene frequency distribution linked to specific traits within socially defined racial groups continentally labeled as African, Asian, European, and Native American. In this way, supposedly value-neutral statistics makes these groups genetically distinct from one another. Continental demarcations are also used as a means for purifying and eliminating the obstacle of group members' genetic heterogeneity. This means, as Fullwiley put it:

> . . . the genetic markers that West Africans, for example, share are separated out from those that they do not share. This sorting takes place through statistical analyses that assess the probability that select alleles will occur at a lower or greater frequency in one of the geographically based, circumscribed populations. The scientific practice of DNA-based group homogenization allows the biologist to perform the task of selectively differentiating between any two "continental populations" of "Africans" and "Europeans," or "Africans" and "Native Americans" and so on.[15]

Examination of the research that many interviewees cite to advance their contention that biogenetic 'races' exist reveals the presence of racial cognitive structures, more than evidence of distinct racial genetic variation, was the basis of their bio-'race' claim. For example, a 2009 study by Kathy Albain and colleagues involving a meta-analysis of data of more than 19,000 patients concerning treatments for a variety of cancers found "that all other/actors being equal, black patients had on average a significantly lower cancer survival rate than whites."[16] The study's authors conclude that unrecognized biological "factors must be contributing to differential survival outcomes by race."[17] The authors attribute cancer survival differences between socially defined racial groups not merely to socioeconomics or access to health care but to genetic differences between the so-called 'races'—implying that a biological category of 'race' exists. With this troublesome claim, policy interventions are deemed ineffective in remedying cancer survival differences since it is a matter of biological differences.

DISABLING DISCOURSE OF RACIAL IMPARTIALITY

Facts are never neutral, and they do not speak for themselves.[18] Facts are subsumed in genomics' racial assumptions then interpreted. So, however useful genomic explanations may be, they will always have certain starting assumptions built in. For example, on the rare occasion when a genomicist points out the social bias of scientific practice, their associates often dismissed them as 'politically correct' pseudo-scientists whose own reasoning is void of scientific rigor.[19] The labeling of critical reflective genomicists as ideologues often occurs very subtly, as illustrated in this informant's explication of the debate between A. W. F. Edwards and Richard Lewontin on interpopulational genetic differences:

> Edwards and Lewontin are both right. Lewontin said that between populations, a fraction of the variance is very small in humans, and this is true, as it should be on the basis of present knowledge from archeology and genetics alike, that the human species is very young. . . . *Lewontin probably hoped, for political reasons,* that it is TRIVIALLY small. . . . In essence, Edwards has objected that it is NOT trivially

> small, because it is enough for reconstructing the tree of human evolution, as we did, and he is obviously right (emphasis mine).[20]

By framing Lewontin as 'political,' the interviewee implies that his genetic variation work is subjectively tainted while Edwards's research reflected the work of an unbiased, objective, truth-seeking scientist. This is a devious construction considering that interviewees' declarations of 'objectivity' are in themselves political acts that imply their unwillingness to struggle with facts that do not fit into their preconceptions. Therefore "objectivity works to repel the attacks of critics, like a kind of [scientific] ethical pepper spray, to ensure some conceptions never make it to the genomic interpreter's table."[21]

Claiming impartiality, then, provided informants with a disabling discourse of 'race,' which obscured how in effect 'objectivity' affirms 'race,' racial thinking, and racism. Such obfuscation is observed in this interviewee's exhortation regarding science and racism: "I don't think racism is a good thing. I think it's a very bad thing. That is my moral position. I don't see any justification in evolution either for or against rac[es]. The study of evolution is not in the business of providing justification for anything."[22] But, genome science is not beyond social, economic, and political forces; rather the idea of scientific 'objectivity,' as I illustrated above, provides researchers with a commonsense rationale that assists in obfuscating racial ideology's influence in interpretations of genetic differences.[23] As a commonsense part of genomic practice, 'race' is used almost reflexively, becoming difficult to dislodge from the conceptual systems of genomics. To instigate the process of removing this barrier genomicists need to acknowledge racial thinking's persistence in genome science in a manner that recognizes genomics' embeddedness in racism and the racialized social structures shaping it.

THE CENTRAL DOGMA'S INSTANTIATION OF 'RACE'

The basic idea of the central dogma of molecular biology is that DNA determines who we are as individuals. This means, as Francis Crick, the coiner of the term, puts it:

> that once "information" has passed into protein it cannot get out again. In more detail, the transfer of information from nucleic acid to nucleic acid, or from nucleic acid to protein may be possible, but transfer from protein to protein, or from protein to nucleic acid is impossible. Information means here the precise determination of sequence, either of bases in the nucleic acid or of amino acid residues in the protein.[24]

Though this static or inflexible view of genetic inheritance is currently confronting a challenge from the new genetic subfield of epigenetics, which posits that changes in the environment can directly alter genotypes, genomicists continue to adhere to facets of genetic determinism in their research. For example, respondents frequently used deterministic thinking in their discourse of about the virtues of personalized genotyping for treating disease:

> I don't want to oversell the work of individualized molecular medicine and human genomes, but that's the direction it's going to go in. You're walking into your doctor's office with your genome on a stick, shoot into the computer, and it's going to read out, potentially, what your variants are. You're going to know from your channel proteins, your different cell signaling proteins, which chemotherapy drugs are more likely to work, which blood pressure controlling drugs are more likely to work, which hormone drugs, are more likely to work.

In this way, interviewees' commonsensically assumed genes are the primary driving force for disease expression. This deterministic outlook is unmistakable in a recent editorial in *European Molecular Biology Organization Reports* on mitochondrial replacement in which authors open with this declaration: "The development of new genetic technology has enabled a genetic diagnosis to be established in an increasing number of patients. That has led to a recognition that mitochondrial diseases are not rare, but one of the most common group of genetic diseases."[25] The authors' claims ignore the fact that even when a mitochondrial DNA mutation is transmitted to offspring this does not automatically mean the child will develop mitochondrial disease. The central dogma's idea that genes determine whether or not we will develop a disease is the locus for sustaining the folk notion that the physical (outward and inner) makeup of a person or group indicates something about their phenotypical essence and propensities.

Respondents' central dogma discourse reveals that unmarked racial assumptions in the central dogma's heuristic strategies (i.e., educated guesses, intuitive judgments, and common sense) racialize human genome knowledge. In short, racial ideology is built into the central dogma. As a mental model, the central dogma as both a representation of genetic knowledge and information-processing uses taken-for-granted typifications and cognitive shortcuts to facilitate effective genome knowledge at the expense of synoptic accuracy.[26] Within the schematic reasoning of the central dogma are the conceptual components through which racial ideology configures genomicists' interpretation of human genome variation. Genomicists rarely depict all the particular details when describing the central dogma process through which DNA alleles encode RNA to translate into a DNA's likeness. The RNA then serves as a template to build proteins expressed as physical composition. In this respect, the central dogma heuristic is a truncated abstract description of genetic sequencing that is instantiated with a more specific and complex socially fashioned description of the genome process. In other words, the central dogma, as a preliminary representation, uses heuristic strategies informed by genomicists' social experiences to define the benchmarks through which genetic differences are understood.[27] This influence is seen in one respondent's unassuming comments regarding the comprehension of disease through genetic data:

> "... if we're looking at groups of people, I'm trying to interpret genetic data, or if I'm looking at a particular person that has a physical finding, and I look at their genetic information as a result of whatever testing been done. And ... I don't know what the particular person's upbringing is, as I'm interpreting that genetic test."

But genomicists know the socially defined racial groups that they defined as being susceptible to the disease under study. For as the respondent explained: "If you're looking for a melanoma gene, you're going to have different incidence of that cancer in people who are from, who live in Africa than people who live in South Dakota." Through prior knowledge regarding the manifestation of disease in particular socially defined racial groupings in specific geographical locations, the central dogma's genetic determinist logic is racialized and fixed in informants' thinking about genetic differences and disease.

Though it is possible to lessen this central dogma heuristic bias through awareness and reflexivity, more often than not informants downplayed or ignored how racialized culture shapes their analyses. Hence when genome researchers employing the central dogma in a more reflexive and critical way reveal that gene reproduction and phenotypes emerge from complex interacting causal processes (i.e., genes, environment, gene and environment, covariance of genes on environment, chance, etc.), unreflexive users are inclined to minimize the fact that phenotypical (e.g., disease susceptibility and drug metabolism) differences are irreducibly complex phenomena. This slanted bias exists in genomics largely due to the fact that researchers in racialized cultures tend to view different disease susceptibilities in social groups as evidence of biological 'races.' But as this respondent observed, though "some races have different frequencies for certain diseases, [w]hether that's genetic or environmental [or both]. . . . depends on the disease." Within such a multifactorial genetic comprehension matrix, racialized informants turn to less complicated familiar ways of knowing to make sense of disease susceptibility differences. This occurs among genomicists through their discourse according to one observant respondent because:

> ... these are scientists. ... When they talk in public ... they say things they know are pretty true. When they talk among themselves they don't know how to articulate [environmental factors] in a scientific way, so they fall back on what they are comfortable with. So by that I mean this. So, if a geneticist goes out in public and says "schizophrenia has a strong genetic component, but the environment also influences

> it." It's probably 16 percent environment, 40 percent genetics, right? Then when that person is in [the public] . . . giving a lecture or is in the laboratory, when somebody says, "Well how do you measure environment?" You know there's a problem because they're not savvy [at evaluating the effects of the environment].

Because most genomicists are largely uninformed about and untrained in measuring environmental influences—social conditions shaped by sociopolitical forces—they tend to excuse not dealing with such factors due to their 'complexity,' which confounds their simplistic determinist explanation about genetics and disease. One respondent asserted that feigning the messiness that complexity induces in genome reasoning is merely a cop-out:

> When you're ignorant . . . you have no idea . . . about imputing that the environment is very complex—which by the way is a separate argument of mine. Some of my colleagues use the word complex to mean it's going to take a lot of money to solve. I have absolutely no idea whether it's complex. I'm completely ignorant. When in fact we do identify a disease's etiology it may be simpler than we suppose. I don't mean it is *simple*. I mean it's simpler than we supposed. So let's not . . . think that this is going to be so complicated that we'll never figure it out. I just don't think we have good ways . . . of looking at the environment, classifying the environment, measuring things.

The idea that genomicists do not "have good ways of looking at environment" is also an excuse since methodologies measuring environmental factors are frequently used in genetic epidemiology in ways that account for how such things as limited access to health care and socioeconomic status inform human genetics.

Despite the existence of these methodologies, respondents were woefully ignorant of them. Thus the disregard of 'race' conceptions in central dogma problem solving compelled most to insist like this individual:

> . . . if somebody self identifies as being Puerto Rican, they've got three genetic ancestries going on and you don't actually know what the susceptibility region of the genome, where you go that whether it was African or Northern European or Asian, Native American. So that's the tenuous part of the link. So right now, yes there are some reasonable grounds to believe that you can have self-identified social construct of race as a kind of a marker, a kind of a predictor for genetic predisposition. I think that's a completely fair statement. But we will get so much better at making those connections when we have more empirical data, just a better foundation.

Although many informants recognized 'race's' infusion in their genomic sense making, they did not hesitate to use it as a stand-in for inferring genetic variation. This, as we have seen, is due to interviewees' lack of reflectivity about the racial undertones in their perceptions and understandings. For this reason, respondents, like this genomicist, routinely asserted:

> I do think it's important to use the information that we have about the variants that are out there, and maybe it would associate with certain populations like Jewish populations, or African American populations. I mean, I think it's important that we do recognize that there are some sequence differences, and . . . that can be very useful for designer drugs.

Note how the comments construe racial and ethnic categories as biomarkers for creating target medication for individuals based on the assumption that people within socially constructed racial groupings are genetically similar. Informants expressed such genetic racialization despite the fact that there is ample opportunity in the central dogma heuristic to instantiate the effect of other factors like gene-environment interaction on variation.

The insinuations of racialized notions in the central dogma's heuristic strategies are so commanding that they confound genomicists' endeavors to explain genetic processes without using essentialistic racial frames and language. This is evident in this individual's statement about the importance of genotyping disease susceptibility:

> For health-related subjects. It's no doubt that ancestry [read: 'race'] related differences, the frequencies of genes and the exposure of genes, we have to take that into account. I'm absolutely convinced from what we've done, others have done . . . that it not helpful to define somebody as red, white, blue, green [when inferring genetic disease risk]. But the genes themselves are a better way of figuring out which genotypes have [a] population risk and whether the genotype itself is the risk factor.

The inscription of 'race' as genes occurs, as I mentioned earlier, when the interviewee uses population, a designation that groups people based on 'geographic' location and 'ancestry,' to indirectly imply that shared inherent genetic variation signifies 'race.' Moreover, respondents, like this individual, spoke of the central dogma's representations of genetic processes as if they are merely reflections of objective biological reality, which racially biased individuals used to either advance or impede human improvement:

> Genetics is a class of information that is potentially quite powerful. And like all powerful forms of information, it can be used for good or for evil if you will, meaning it can be used to advance the state of human civilization, or it could be used to try and push us back into the 14th century. And most people will want to do the good stuff, and some people will have a vested interest in doing the bad stuff with that same information. . . . Most people . . . now, so this is again cultural upbringing. Most people are mostly good. But . . . you can't deny the

> existence of the dark forces in our dualistic world. And the dark forces will use genetics to whatever their ends are, or attempt to. And so that's why genetic information has to be controlled by acceptable uses and rules of acceptable conduct.

Contrary to the interviewee's assertion about 'objective' genetic information, the search for genome difference is driven in part by normative racial thinking that melds into their reasoning as unquestioned ways of knowing. As such, genomics assigning "someone to a race" based on 'geography,' 'ancestry' or 'population' attributes primary importance to these designators over all other inducers of human genetic variance.[28] This tendency is observable particularly among prostate cancer geneticists whose work, like most human genome variation research, is predicated upon finding genetic distinction to explain why certain socially defined racial groups are at greater risk for the disease. Although there are numerous prostate cancer studies that include 'race' as a variable or category in their analyses that find no genetic differences among groups, these investigations are usually not reported in journals, whereas those that report finding differences are published and hyped.[29] As a result, genomicists reading these published works come to accept genetic differences between 'races' as conventional wisdom even though the majority of research finds no group differences.

The central dogma's genetic determinist focus, coupled with the societal importance placed on finding racial biological differences, to explain varying disease susceptibility rates between groups moves respondents to de-emphasize the significance of social determinants, broadly defined as social, economic, and political resources and structures influencing the disease development. Because genomicists' racialized sensibilities incline them to accept the central dogma's deterministic message as a truism their research tends to devalue the environment's impact on the human genome. For instance, though variability in drug response among people is multifactorial, including environmental, genetic, and disease determinants that affect disposition (absorption, distribution, metabolism, and excretion), study after study[30] ignores the effects of social elements to focus on 'finding' a genetic basis for explaining why marginalized socially defined racial groups are, on average, more likely than 'whites' to be slow metabolizers of certain drugs like antidepressants. This genetic bias is evident in lung cancer studies of the drug gefitinib (Iressa), which reported that Asians with the epidermal growth factor receptor (EGFR) genetic mutation had a better response to the drug. However, in subsequent studies when researchers controlled for other factors the difference in response among individuals with and without the receptors disappeared. Moreover, in successive clinical studies of gefitinib, the U.S. Food and Drug Administration's found the drug had no survival benefit.[31] These findings suggest that differences in drug response vary widely between and within socially defined racial groups. Nevertheless, most oncogeneticists remain steadfast in their search for bio-'race' related genetic causes of cancer.[32] Interview data repeatedly reveals that genomicists' unwavering search is due in no small part to racialized ways of knowing they use to make sense of the variation. However, this way of understanding differences in disease susceptibility rates forces all other variation consideration into the background.[33]

Despite the growing understanding of the importance of social determinants for disease, genomic research is only now beginning to investigate social determinants' influence on the human genome. Addressing the social determinants involved in disease risk can provide researchers with the conceptual frame needed to alter the central dogma to account for how people's embodiment of social hierarchy may increase and decrease disease incidences.[34] One informant explained that this is the approach epigenetics takes when studying disease variation between groups:

> I just want people to know that there's something over and above just the genetic sequence itself that can influence your risk of disease. And that this is something that can be influenced by an environment and is potentially reversible so that you may . . . for example, if you smoke, we know that induces DNA damage. We also know it induces epigenetic changes.
>
> And so *the question is can you reverse some of these epigenetic changes with some sort of intervention that might allow you to decrease the components of the disease susceptibility* that came from those changes. It's a very exciting area of potential intervention. We haven't got there yet, but I think it's great legs for the future.

Though informants tout epigenetics/epigenomics as a more holistic investigatory approach for understanding variation in disease susceptibility, its strategy, as I alluded to earlier, remains focused on changing genes not the environment that alters the genome. As Paul Farmer and colleagues observed, such biological framing in disease etiology studies leads to an exclusive focus on molecular-level phenomena that contribute to the "desocialization" of genome inquiry: "a tendency to ask only biological questions about what are in fact *biosocial* phenomena."[35]

Given that the central dogma's heuristic strategies are a conduit for racial ideology's insinuation into genetic reasoning, some evolutionary biologists, like this interviewee, contend that we cannot deny the biological basis of 'race' simply because "human races do not exist in the sense that biologists apply the term to animals."[36] Though such a simplistic contention is not representative of genomic thinking, other geneticists do stress, albeit more implicitly, the importance of 'race' as molecular reality by contending "a decade or more of population genetics research . . . document[s] biological differences between the races."[37] But what the research cited more accurately documents is the racially modulated alignment of genomicists whose study designs and genetic interpretations are constituted in ways that are consistent with their racial socialization. Since genomic practitioners are racially situated they think and perceive of genetic differences between socially defined racial groups as biologically racial and 'obvious,' they are less likely to see the clinical error in developing treatments targeted at particular socially defined racial groups. This is not to say that 'race' specific drug development is not driven, as historian Jonathan Kahn correctly points out, by multinational pharmaceutical companies seeking to establish niche markets for their 'medicine.'[38] Desire for medicinal markets is what motivated NitroMed to petition the FDA for approval of the United States' first 'race'-specific drug BiDil, supposedly tailored made to treat congestive heart failure in African Americans. After gaining approval the drug failed spectacularly because there was no scientific evidence that racial physical features have anything to do with how BiDil works. This became clear when it was disclosed NitroMed's African American Heart Failure Trial (A-HeFT) enrolled only self-identified African Americans. With no comparison population, no legitimate claims could be made that BiDil performed differently or better in African Americans than in anyone else.[39] Through A-HeFT's problematic design, 'race' as biology took center stage in the BiDil trials and FDA approval process even though it is not a coherent genetic concept. In spite of the fact that it is well known in genomics that 'race' statistically explains only small portion of genetic variations, most genomicists continue to use racial identity as a guide for disease susceptibility.

Preoccupation with finding bioracial rather than social determinants diverts genomicists' attention away from investigating what one respondent considered the primary question of genomics: "How did [a gene] come to be like *this* in [this human group] while it is like *that* in some other group, and yet this is basically the same gene or the same molecule found in all groups?"[40] The basis for the question is apparent in sickle cell disease research that once erroneously approached this malaise as a 'black disease,' thinking it to be a gene distinction in people of African descent. More comprehensive research revealed sickle cell to be a disease of people originating in areas with high incidences of malaria. Some of these populations are African and some are not. The four distinct sickle cell linked genes are found among people scattered throughout high malaria geographical zones. Whether or not these genes are expressed depends on an individual's environmental, not racial, location. By disregarding the symmetry between elements of the central dogma and racialized social reality, intellectual space is opened in genomics to construct essentialist elaborations (such as 'race'-specific therapies) that come to reinforce each other and create the impression of evidence when in fact there is little or none.[41]

Racial thought's infusion in respondents' 'objectivity' discourse and heuristic practice suggests there are no pure genomic 'facts,' only genetic realities couched in one sociopolitically constructed conceptual system or another. The implicit racial bias operating within informants' genomic conceptual field did not require them to devote attention to its expression; instead it worked quickly and unknowingly through 'objectivity' discourse and the deterministic heuristic strategies of the central dogma. For this reason the relationship between genomic observation and racial interpretive assumption are hopelessly circular, with interpretation generating the observations that supposedly support them. Genomic knowledge production is guided by antecedent racial values and ideas that contribute to the motivations and conceptual outlook of genomicists. For it is through mechanistic and reductive molecular assumptions of the central dogma that fundamental and deeply seated racial ideas about differences are constituted as bio-'race.' Given this, racial ideology in genomics, intentional or not, simply takes culturally pervasive, but highly questionable, beliefs in inherent differences between socially defined racial groups and imprints them in the central dogma. In so doing, the power of genomics to describe human genetic variation reinforces racist biological determinist beliefs and practices, despite genomicists' good intentions.[42]

NOTES

1. Haraway, "Situated Knowledges: The Science Question in Feminism and the Privilege of Partial Perspective."
2. Haraway, "Situated Knowledges: The Science Question in Feminism and the Privilege of Partial Perspective."
3. Mannheim, *Ideology and Utopia*, 50.
4. Gould, *The Lying Stones of Marrakech: Penultimate Reflections in Natural History*, 104–105.
5. Haraway, "Situated Knowledges: The Science Question in Feminism and the Privilege of Partial Perspective."
6. Haraway, "Situated Knowledges: The Science Question in Feminism and the Privilege of Partial Perspective."
7. MacKellar, "Ethics and Genetics of Human Behavior."
8. Fee, "Racializing Narratives: Obesity, Diabetes and the 'Aboriginal' Thrifty Genotype."
9. Boggs, *Gramsci's Marxism*, 31.
10. Khan, "Interview with Bruce Lahn."
11. Balter, "Brain Man Makes Waves with Claims of Recent Human Evolution," 1871, 1873.
12. Abu El-Haj, "The Genetic Reinscription of Race."
13. Abu El-Haj, "The Genetic Reinscription of Race."
14. Fullwiley, "The Biologistical Construction of Race: Admixture Technology and the New Genetic Medicine."
15. Fullwiley, "The 'Contemporary Synthesis': When Political Inclusive Science Relies on Biological Notions of Race."

16. Catherine Elton, "Why Racial Profiling Persists in Medical Research," *Time Magazine*, August 22, 2009. Accessed June 2015, http://content.time.com/time/health/article/0,8599,1916755,00.html; Albain et al., "Racial Disparities in Cancer Survival among Randomized Clinical Trials Patients of the Southwest Oncology Group," 984.
17. Albain et al., "Racial Disparities in Cancer Survival among Randomized Clinical Trials Patients of the Southwest Oncology Group," 990.
18. Clouser, *The Myth of Religious Neutrality: An Essay on the Hidden Role of Religious Belief in Theories*, 26–27.
19. Fausto-Sterling, "Refashioning Race: DNA and the Politics of Health Care"; Graves, *The Emperor's New Clothes: Biological Theories of Race at the End of the Millennium*, 2, 161.
20. Khan, "Interview with Luigi Luca Cavalli-Sforza."
21. Quote from Gladstone, *The Influencing Machine*, 102; DeMar, "The Myth of Objectivity."
22. Miele, "Darwin's Dangerous Disciples: An Interview with Richard Dawkins."
23. Pierre Bourdieu, *Science of Science and Reflexivity*.
24. Crick, "On Protein Synthesis."
25. Hebert and Turnbull, "Mitochondrial Replacement to Prevent the Transmission of Mitochondrial DNA Disease," 539.
26. DiMaggio, "Culture and Cognition."
27. Bourdieu, *The Logic of Practice*.
28. Lock, "The Tempering of Medical Anthropology: Troubling Natural Categories."
29. McDowell et al., "Systematic Review and Meta-analysis of Ethnic Differences in Risks of Adverse Reactions to Drugs used in Cardiovascular Medicine."
30. Lehrman, "Race, Genes, and Illness."
31. National Cancer Institute, "FDA Approval of Gefitinib," National Institutes of Health.
32. Lehrman, "Race, Genes, and Illness."
33. Lock, "The Tempering of Medical Anthropology: Troubling Natural Categories."
34. Baker et al., "Addressing Social Determinants of Health Inequities: Learning From Doing."
35. Farmer et al., "Structural Violence and Clinical Medicine."
36. Malik, "Why Both Sides are Wrong."
37. Risch et al., "Categorization of Humans in Biomedical Research: Genes, Race and Disease."
38. Kahn, "How a Drug Becomes 'Ethnic': Law, Commerce, and the Production of Racial Categories in Medicine."
39. Kahn, *Race in a Bottle: The Story of BiDil and Racialized Medicine in a Post-Genomic Age*.
40. National Academy of Achievement, "Interview: Ernst Mayer: The Darwin of the 20th Century."
41. Kaufman, "The Anatomy of a Medical Myth."
42. Spanier, *Im/partial Science: Gender Ideology in Molecular Biology*.

FIVE

'Bad Science' Discourse as Covering for Racial Thinking

Interviewees tended to use the 'bad' or 'good' science discourse when speaking about the therapeutic benefits of human genome variation research. Though there is considerable discussion in genomics about 'good' and 'bad' genome science, there is very little in-depth critical engagement with what this means except in the general sense of 'bad' genomics not being evidence-based. In this chapter, I examine how the 'good/bad science' frame obscures the racial logic embedded in genomics' methodologies and endeavor to create personalized genomic medicine. Genomicists, as already discussed, are socialized racial individuals whose racialized disposition lends validity to biological racial ideas. Hence analyzing the racial ideology within the 'good/bad science' refrain, and understanding the interests served by this particular rhetorical approach, is of fundamental importance for critically engaging racism's impact in human genomic knowledge production and application. In what follows, I investigate how the 'good/bad' genome science rhetoric is constructed, employed, and consequential for people's health.

The starting point for my examination, like that of preceding chapters, is with the social constructionist view that genomic sense making rests not with scientific 'objectivity,' but in the context of genomicists' racialized cultural values which equip them with commonsense racial ways of knowing and explaining genomic processes. We have seen interviewees, often unwittingly, safeguarding their scientific cognitive authority using the dispassionate scientist rhetorical tactic to distance themselves and genome science from the racism in which both are embedded. Similarly, interviewees detach themselves from scientific racism by differentiating their therapeutic research concerns from 'bad' genome science's harmful preoccupation with finding racial biological differences to explain susceptibility variation between socially defined racial groups. However, the 'good' genome science mind-set invariably hinges on separating science from society in a manner that overemphasizes the importance of genetics in explaining disease susceptibility differences. In so doing, the 'good/bad science' division reifies 'objectivity' in genomics in contradistinction to the fact the discipline as a human creation is intimately entwined in cultural values. This makes it difficult for genomicists to reason their way out of the racial thinking that they did not reason themselves into.[1] By ignoring this cultural link the 'good' genome science discourse enables genomicists to embody genomic knowledge with racial notions to constitute 'race' as slight genetic variant differences between socially defined racial groups. Racial truth is a product of racialized culture which produces it through multiple social structures.[2] In other words, bioracial truths are sustained by scientific discourse and institutions, and constantly reinforced (and redefined) through the education system, the media, and continuously changing but racially congruent political and economic ideologies. In this regard, genomicists' human genome variation explanations are not absolute 'objective' descriptions, but rather culturally cultivated scientific interpretations of genetic difference. Racism and human beings do not exist apart from one another; they are in constant interaction with each other. The 'race' concept, as a cultural production of human action, acts as a boundary enabler that constricts genomicists' relative capacities to know. This is exemplified in this informant's observation about the concrete consequences that racial linguistics have for understanding genetics:

> You know . . . there has been an evolution in the language that we use to describe persons of African ancestry. You know, when I grew up the, you know, you'd still hear the term Negro, and nobody uses that anymore. And then for quite some time you know, it was appropriate to refer to people of African ancestry as blacks. And then it became, it was suggested that African American was a better usage. And so . . . when I'm presenting to an audience, presenting scientific findings to an audience I tend to use the term African American. I was severely taken to task by somebody in the audience who said, "So, Charlize Theron [well-known actress] grew up in South Africa. So she's an African American. . . ." [T]he problem is that in the English language the colloquial English language does not have the appropriate terms that reflect genetic reality.

Another observant respondent added: "[The race concept] is so strongly connected with American institutions that you cannot . . . it's very very hard . . . to take it out of the discussion. It's very very difficult. And so for folks [genomicists] who really care less about whether race is real or not. I mean, they're going to continue to use it." As a commonsense means for understanding human differences, racial ideology represses, censors, masks, and conceals its operation in genomic variation knowledge; yet it frames the methods for collecting and interpreting genetic evidence. Genomic knowledge, like any knowledge in racialized societies, emanates from racist power dynamics.[3] As such, I deconstruct the 'good/bad science' discourse to reveal its role in obscuring the racial ideology at work in genomics. In so doing, I illuminate how genomicists, as racially socialized and embodied individuals, construe genetic processes that do not fit their bioracial ways of knowing as unimportant or configure them to correspond to their racialized common sense.

RACIALIZED METHODOLOGICAL ASSUMPTIONS

In interviewees' discussion of genomic variation, being 'objective' thus a 'good' scientist meant applying the rules of science fairly, and not letting bias creep into the production of genomic knowledge. Their assumption was that doing 'good' science through the verification of objective facts, correction of mistakes, and reliable reporting of findings always led to the objective truth about genome reality. But this is usually not the case, according to researcher John Ioannidis, because all scientific studies are biased. "Sometimes they [are] overtly biased. Sometimes it [is] difficult to see the bias, but it is there."[4] Genomicists researching genetic variation head into their studies looking for certain results and are finding them. Though the scientific process involves being rigorous about separating out what is factual from what researchers merely wish to be true, it is relatively easy to misconstrue facts unintentionally or unconsciously due to genomicists' internalized racial understandings informing the questions they ask, how they design their study, the selection of individuals and groups to study, and their analysis of genetic data.[5] An outcome that is seen in this interviewee's explanation for becoming a population geneticist:

> Interviewee: [O]ne cannot do high quality research on any kind of human genetics without thinking about geographical ancestry of the population of humans that you're studying. . . . But here in the United States, basically what we're talking about is [that] you need to know if somebody's white or black before you can intelligently approach including them as participants in human genetic research. . . . And again one of the things that attracted me to the field was the challenge of trying to be sensitive to those issues and to do *good science*, all at the same time.
> Interviewer: What do you mean by 'good science' though? Explain that . . .
> Interviewee: Well, what I mean is, for example, something that I wish I could say I helped develop, but not smart enough to have developed these [new] methods . . . [that] are reasonably precise for genotyping. . . . [I]n this particular study, we're only going to recruit white people. Or we're going to recruit black people. Or we're going to in-

> clude this particular group in this statistical analysis because we don't want to be mixing groups *because we know that* that's going to introduce this league of variants in the genetic markers.

The informant brings to the study preconceptions about the correlation between 'race' and genetic composition that shapes the aforementioned genome project's design and analysis. Such bias impedes interviewees' ability to see what there is to see, given that they, like everyone else, have to *learn* to see the genomic facts that are there through their particular culturally distorted conceptual lens.[6] As standpoint theory observes, since there are multiple ways of interpreting genome variation reality, "seeing things one way means not see them another way."[7] Biologist Fred Grinnell conveys this understanding this way:

> Each person's previous experiences . . . le[a]d to the development of particular concepts of things, which . . . influence what objects can be seen and what it will appear to be. As a consequence, it is not unusual for two investigators to disagree about their observations if the investigators looking at the data according to different conceptual frameworks. Resolution of such conflict requires that the investigators clarify for each their concept[ual frame].[8]

In other words, though genomicists share scientific background assumptions about the genome, their differing unexamined racial suppositions configure how they perceive and frame human genome variation. A *small* number of interviewees, like this individual, recognized this influence:

> Again, they (researchers) are wrong. I'm going to tell you why they are wrong, because their methodologies . . . are making assumptions about the nature of population [genetic] structures. One of the algorithms they use is called 'Structure.' What Structure does is it takes genetic data from human DNA sequences and then clusters individuals into groups. The way it does that clustering starts off with the assumption that there are underlying groups in the data. So, in other words, it's an algorithm designed to create groups assuming the groups exist. The algorithm itself—which is one their main tools—starts out with a platonic assumption of the underlying existence of separate racial groups. What they are doing . . . then is determining which individuals belong to which racial groups. They called them clusters but that is exactly what they are doing.

Among the few informants who were aware of 'good' science's racial bias, there was skepticism about convincing their colleagues to account for racial ideology's operation within genomic infrastructure and design. Such averseness is displayed in this interviewee's story about writing a grant with some colleagues:

> . . . [who] had no idea of my position on race when we started writing th[e] grant. . . . And so . . . they sent me a draft... frame[d as] racial differences. . . . And so I went through th[e] grant and reworded and

> restructured, re[wrote] . . . And then I said, "Look, you know, we can't—obviously, you know, we're trying to leverage the grant on differences on the disparity in terms of drug response—but... this is not a racialized model (inaudible) . . ." And so, it was educational . . . for them. . . . [S]o, by the [time we completed] writing [the] grant . . . they were well educated, I think. And I don't think that they're going to come back, at least to me, with another piece, another document [framed] like that.
>
> So, I'm curious to see . . . when they write their papers if they're going to still adhere [to] how [they originally] structured [the] grant or if they're going to go back to th[e] racial model. I mean . . . the title of the grant, *Racial Differences in Drug Response*. This makes no sense. [W]e know that there are differences in . . . individual differences in drug response and so we should design this in a much more sophisticated way because it's not just about black/white genetic differences. And they just can't . . . they don't want to give it up. It's like an addiction, you know?

One observant respondent confirmed my earlier contention that genomicists are hesitant to construct a more involved genetic variation methodology due in part to their lack of training in accounting for variance induced by nongenetic factors. The respondent maintained that though

> socially defined races differ in disease profiles, the question is why? . . . [T]his is basically a quantitative genetics[9] problem. . . . [A] lot of [genomicists] don't understand quantitative genetics either. Any phenotype—a disease is a phenotype—is produced by five different sources of possible variants. The first variance is due to genes and that is what most of these folks focus on. That is not the entire equation; then you add the variance due to environment. Say if you have two trees, one of which is growing out of cliff in a really windy environment . . . you will see over time the tree will bend over—literally bend over—and then growing in a place where there is no wind, the same [tree] will grow straight. If you were to take the two sets of [tree] seeds and put them in different environments, the phenotypes switch. So, it's clearly an effect of the windy slope causing the bending of the tree, not genes.

The respondent continues:

> . . . [T]hen there is variance due to genes by environmental interaction because different genotypes respond to different environments differently. . . . And then [there] is the most damning of these kinds of genetical explanations disease—the covariance of genes in environment. What that means is that you may find one genotype preferentially in a good environment—like European, wealthy Americans—they have European genes and they are preferentially in a good environment. Then you have another genotype—persons of African descent—who are predominantly in poor environment. . . . So the difference in the phenotypes you measure between these two groups is not due to difference in genes but . . . the distribution of the genotypes in good

> versus bad environment. And finally . . . there is variance due to error. . . . Assuming you have competent researchers . . . the error/unit measurement is [will] be very small. These are the large sources of variance in complex traits. So if you put all five of [these variances] together that's how you explain a disease differential.

But as the respondent explains genomicists tend to measure "genetic differences then . . . assume, 'Oh, we've measured a genetic differences and one group is sicker than the other; we assume that the gene is responsible for the second group being sick.'" Because a vast majority of genetic researchers do not know how to measure other variance types they engage in what the respondent termed 'genomic reduction' which encourages them to be 'intellectually' fictitious about its inclusion in their methodological assumptions.

Such reduction is a by-product of the naturalizing genetic axiom that 'like begets like' which encourages many genomicists to stratify social groups based on the appearance of disease or genetic mutation in individuals in the same socially defined racial groupings to suggest genetic similarity. Since the implicit assumption of this logic is that genotypes are inflexible structures—a central dogma tenet—genomicists search for similar genetic patterns within socially defined racial groups to explain group, not individual, variability in disease. Yes, genetic factors influence the incidence of disease but so too do a broad of range of social conditions. The quest to find genetic susceptibility diminishes the importance of accounting for the influence of other nongenetic variance in discerning disease susceptibility differences among groups. Genomicists' disregard or downplaying of such factors predispose them to perceive and present nongenetic factors as intractable elements that are too 'complex' for existing genomic methodology to decipher. Though genomicists generally understand that a more complete explanation for disease susceptibility difference is found in the interplay between biology and nongenetic determinants, their singular focus in genomics, as the respondent above explained, is its genetic underpinnings. For example, genetic sequence differences are strongly correlated with high risk for diabetes in different social groups, suggesting explanations for susceptibility differences are genetic. But, the number of people in the United States with diabetes has doubled in the last twenty years. Evolutionary biologist Joseph Graves correctly observes this increase simply cannot be the result of

> massive changes in the U. S. gene pool. During this period, it has been shown that race, sex, obesity, and age are consistent predictors of the prevalence of diabetes. With regard to race, the question must be asked, is it the genetic or social aspect (or their interaction) that account for the consistent link between the race variable and complex disease?[10]

This is line of inquiry, however, is sparse given that genomics' methodologies are generally driven by implicit assumptions regarding the pertinence of genetic aspects of 'race' for explaining disease susceptibility variation among groups and individuals when "there is no reason to believe this."[11]

Genome variation research's embeddedness in operational assumptions makes it difficult for genomicists to ascertain that it is not merely a question of 'good' or 'bad' genome science, but rather how they uncritically go about their work. Though there is no doubt that awareness about racial influence in genomic methodology has risen in the genomic community as a result of a small number critical studies and better trained genomicists, most researchers are still relatively ignorant about the impact of racism and its associated ideologies on them and their studies. This is so according to this informant because:

> When we say that science invented this thing called race—yeah it did but these modern [scientists] doing . . . genomic research don't know the narrative of how science invented this thing called race and for what purpose. So their notions of . . . race is at the biological level are compromised. They don't understand the difference between the social definition and the biological ones. So [you never see] one of their papers . . . saying, "human biological races are defined by . . ." you'll never see that. They don't know . . .

This inability to grasp how the supposedly 'good' science of racialized human genomic variation research has potentially deleterious outcomes for marginalized socially defined racial groups' health and treatment in the new genomic subfield of *precision* or *personalized genomic medicine*. In this field, understandings of individuals' genetic health remain grounded in eugenic assumptions about the genetic well-being of their socially defined racial groups.

'PERSONALIZED/INDIVIDUALIZED GENOMIC MEDICINE'

Similar to 'positive' eugenics, what is now termed personalized or individualized genomic medicine uses gene therapies and other genetic interventions in an effort to ensure people are without disease so that they can live longer and healthier lives. The 'good' science of personalized genomics, however, differs from 'positive' eugenics in that its central concern is the welfare of the individual not the group. 'Positive' eugenics conceived of health as an outcome of an entire socially defined racial group's genetic fitness, consequently intermixing of these groups with a supposedly 'fit' socially defined group (aka: middle and upper class 'white' Euro-Americans) was thought to degrade their 'gene pool.' Given this, proponents of 'positive' eugenics moved to control—"by means of altruism and sentimental charity"—the reproduction of groups and individuals designated as genetically 'unfit.'[12] According to Nadia Abu El-Haj, the categories of "altruism and sentimental charity" comprised support for the medical care and intervention that personalized genomic medicine promotes.[13] Unlike eugenics, personalized genomic medicine pushes for the medical care of the genetically 'unfit' by touting the importance of genomics for improving genetically 'imperfect' individuals' health.[14] An excitement evident in this informant's comments:

> So within maybe five or ten years, there will be an increasingly large number of databases which link genomic information—all the information in someone's genome—with other information about them, maybe medical tests or, in some cases, their whole medical records. So, I think we have to think about the world where that will be the norm. There will be tens, hundreds of thousands, or even millions of people with genomic data and health care data in those large databases. And the challenge for us is to extract the information from those databases to learn how particular changes in someone's DNA might make them more likely to react well to this drug or suffer if they have that kind of treatment rather than another.[15]

Geneticist Kenneth Offit reminds us that rather than constituting a new paradigm, personalized genomic medicine is an incremental development that emerged from several decades of scientific discovery and clinical translation (e.g., genetic counseling, genetic profiles of disease) in human genetics.[16] Personalized genomic medicine uses information about a person's genes, proteins, and to a lesser extent environment, to prevent, diagnose, and treat disease.[17] The phrase 'prevention,' rather than 'reproduction control,' is emphasized by genomicists to present personalized genomics as a 'good science' that allows for early intervention via diagnostic test for disease risk. In a word, researchers hype the importance of genomic tests for disease and drug response in ways that allow doctors to categorize patients to [ensure] appropriate treatment, limit bad outcomes, improve care, and lower cost.[18] This less curative, more 'preventive' genome science is still centered in the eugenic vision of a biologically engineered society. 'Prevention' is possible because technological advances are providing genomicists with the ability to manipulate individual genes. With this capacity, informants like this one expressed their belief that medical care would be more effective:

> I think that if you could genotype everyone and know that it's a specific genotype, it affects whether medicine would work. [It] makes a lot of sense to personalize medicine, because a lot of what we do in medicine is a waste. So, we give medicines at the wrong dose for the wrong person, or too much, so too little a dose doesn't get you an effect. If I give the same hundred people the same dose, they're all going to have varied effects. Some are not going to respond, some are going to be toxic responders, and some are going to [have] an average response.

But genomicists' conjectures about an individual's dose responsiveness are grounded in ideas about the person's racial composition. As this interviewee explained,

> When you look at it from a sense of a scientist . . . [we] look at what is causing the disease. . . . Because a mutation in a gene would influence a certain enzyme and protein, regardless of what color you are. That's what is going to benefit, not whether you're black, white, Chinese or whatever. I think . . . scientists we should focus on individual differences . . . not population groups. [W]e have a large group of scientists now looking more into ancestry genomics, which is looking at how each individual may be different—you may have a certain percent of African American, a certain percent of Caucasian, especially African Americans . . . , you know they're a mixture of populations. So, they're doing a lot of whole genome [wide] studies. And trying to make associations between whole genome associations and . . . diseases. Now you may get some data out of that that may be helpful, *but when you look at that type of analysis, [it's] based on the profile and ancestry . . . of those individuals.* [W]hen you convert that down to black and white, you won't be able to [fully grasp what's going on].

Genome wide association studies' (GWAS) 'genetic standpoint' for identifying genes association with disease assesses and adjusts for ancestry difference susceptibility in ways that easily interpreted as 'race' variance. Because genomicists' methods are informed by techniques devised by population geneticists, Joan Fujimura and her colleagues contend, they sometimes use population notions that specify that genetic variation groups are human groups that have shared ancestry.[19] It is through the notion of ancestry that genomicists infer shared ancestry and group individuals with similar single nucleotide polymorphism (SNP)[20] variation into a 'race.'

The relationship between personalized genomic medicine and 'race' is intricate and difficult to disentangle as demonstrated in this interviewee's comments about overcoming this problem:

> The idea of prescribing drugs based on genotype exists anyway, even if there was no discussion of race, right? So for drugs such as warfarin there are well known genetic variants, and I don't know what the racial breakup of these genetic variants would make you prescribe warfarin differently even within white European populations. Again, I'm not familiar with what that is in the blacks. The idea that we want to improve the performance of drugs by using genetics exists before we add any race to the equation. It exists as the idea that this could be helpful for getting the right drugs to the people that need it the most, and I would hate for that message to be lost in a discussion of race.

But as Fujimura and her colleagues show in their work on the operationalization of populations as biomedical research categories, individuals in genotype dosing groups indicate 'race' remains an important organizing concept in their analysis.[21] It became clear, as this interviewee carried on with his discussion, that 'race' is an important means for him to make sense of genetic disease susceptibility.

> If you look at something such as say high blood pressure, hypertension as it goes through a lifespan and you look at the trajectory as we would call it of hypertension increasing as individuals get older, in different races you could see how that as an example might play a role. In other words, I think a clinician might be doing a good job if they had a different threshold for African Americans and for Europeans for when they're going to start anti-hypertension medication.

Though 'race' is clearly presented as an organizing concept for understanding disease, the interviewee attempts to deflect its importance in genomics when speaking about how environments inform gene expression.

> 9P21 is a chromosomal region that has been shown to play a role in cardiovascular disease. . . . And we've shown that it only plays a role in cardiovascular disease if you don't eat enough fruits and vegetables. That's in fact an area of great interest to me, is what we call gene-environment interactions.

Similar to other respondents in the study, after acknowledging the importance of environment in genetic variance and disease susceptibility, the interviewee quickly pivots back to trumping the importance of group genetic variance analysis in determining individual disease risk. This shift is obvious in the interviewee's subsequent remarks:

> We have something in genetics called population stratification and you're suggesting . . . population stratification is actually a problem for us, and so if we take Europeans and Japanese and we decide to analyze them all together and forget their ethnicity, we won't see the truth. [W]e won't be able to run appropriate statistics, right? It's sort of . . . you can even go back to . . . the sickle cell anemia genetic variation, right? I mean if you want to talk about well how likely is this disease to show up and we analyze everybody in the United States, we're going to come to a different conclusion than if we analyzed the African American community, right? And I think a lot of information will be lost if we don't present the analysis of the frequency of that genetic variant in the African American community.

In the more succinct words of another respondent:

> You have to let the biology tell you what the population history of any given individual is. And you do that by sampling polymorphism genome wide . . . using very sophisticated techniques like (inaudible) analysis to understand the genetic structure of that individual. . . . [I]t's only when you do that, that individualized medicine starts to make sense.

Though the goal of personalized genomic medicine is to establish how variations in a person's genetic profile, rather than their racial genetic profile, impact disease susceptibility and drug response, the basis for creating an individualized genomic treatment is grounded in racial group differences. This is the case in part because as this informant put it:

> If you understand it in a group where you can actually study it, you can get more information from that group because the percentage of people might be higher. And then you can use that information in the more general sense to test hypotheses in people from other groups in the population. So, I think it can be very convenient to study a group where there's a particular disorder, that is very prevalent because it gives you more information; it gives you more data points to actually understand the basis of that disorder in that particular group. And then you can use that information to address whether or not the same mechanisms are present across the population, where the disorder might still be very important, but might appear less frequently.

Another informant explains the concern with racial genetics in more precise terms:

> If I knew that race expedited who would get [a] drug, then it is important. . . . So if there's a point behind it, then I think it's worthwhile. . . . [Say], if you're from a specific place in Mexico . . . and I know that a group of people traditionally have married each other, and I know there's a particular gene variant in there, then it would be very important for me to know . . . where that person is from. Watch out giving them a specific medicine because there's a high likelihood that they're going to have a bad effect from it.

In this way 'race' gets implicated in the 'good' science of personalized genomics to propagate research making gross correlations between 'race' and genetic variation.

This all means that scientists utilizing personalized genomic analyses in the context of systemic racial oppression need to be vigilant about racial ideology's incursion into their work. Unrecognized racial sentiments encourage genomic investigators to over-hype the importance of genes for disease prevention. To a great extent, this occurs in genomics because genetic evidence is analyzed as standing alone and not in the context of other facts.[22] According to sociologist Troy Duster, this proclivity fuels

> a persistent search for 'hard data,' for a biological or biochemical explanation for homelessness, mental retardation and mental illness, alcoholism and drug use, even unemployment, violent and abusive behavior. The change has come in a thick disguise of the old concerns, the promise of untold health benefits and the lessening of human suffering.[23]

Racial ideology's endurance within genomics is emblematic of the fact that genomics is rooted in a racialized social climate where many people are socialized to understand human disease and talents as outcomes of the socially defined racial groups' genetic composition to which individuals are assigned. There is no clean break with racial common sense when it comes to personalized genomic medicine. Because the 'race' concept is inscribed genomics, it is presented in both the objective organizations of genomics and the minds of genomicists, making division between 'good' and 'bad' in regards to racialization moot. To paraphrase sociologist Pierre Bourdieu, "the racial pre-constructed is everywhere," thus genomicists working within academic and commercial laboratories are always submersed in racial preconstructions.[24] In view of this, genomicists are burdened with the task of knowing human genomics disembodied from the social world that provides them the language to interpret the genome.[25] This is essentially an impossible feat given genomicists' deep internalization of racial ideology.

RACIALIZED COMMERCIALISM'S CHALLENGE TO INDIVIDUALIZE GENOMIC MEDICINE

Conjoined with racial ideology in informing how genomicists conceptualize personalized genomic medicine is the corporate pressure to sort disease and other aliments into groups in order to maximize profit for genomic treatments and tests. In genetics there is a long history of researchers being divided into 'lumpers' and 'splitters'; lumpers tend to assign individuals with similar genetic variation into rigorously defined categories while splitters takes precise definitions, and create new categories to classify genetic samples that differ. Though advances in genetics and biomarkers favor 'splitters' over 'lumpers,' biomedical corporations lean towards a 'lumper' strategy given their profit aims. This corporate preference enhances bio-'race's' importance in genomics according to this respondent who argued that racial ideologies omnipresent means that 'race' is

> . . . not going anywhere. Not in the United States, it's not going anywhere. So, everybody's talking about individualized medicine [in] the mid-90s or early 2000. . . . And . . . that . . . model is a very different model from how pharmaceutical companies have been operating since they started. Their model is develop a drug. Sell it to as many people . . . as possible. [I]t takes a hell of a lot of money to get that one drug approved. . . . So, you get that one drug that works for many people [so] you market that to everybody. Now, we're [scientists] saying that drugs should be developed for the individual, different versions of th[e] drug and we're going to have to market not to one group, but to many different groups. That model economically isn't . . . the business model for pharmaceutical companies. . . .

> So, who's espousing individualized medicine? It's really the scientists. [P]harmaceutical companies talk about [personalized medicine] because that's what science . . . you know, [because] they want to be up on the science. But that's not their business model. And so, how [do] they achieve [their model]? Well, [they say] let's meet 'em halfway. That's why I [call it] race-based medicine. Let's meet them halfway. So, instead of [focus]ing [on], 'the individual,' we're going to [concentrate on] 'the racial group,' right? And so, [pharmaceutical companies] see which drugs [a]re more efficient for blacks, or whites, or Asians, or Hispanics. And then, [you get] drug[s] . . . like BiDil which actually improved [the health of some] heart failure patients who [a]re black. "Let's market this specific to African Americans." They put a hell of a lot of money in that . . . whole clinical trial and then the lobbying of the African-American organizations. Then presenting it to the FDA and they got it approved. [This drug marked a] movement in the direction towards individualized medicine, but there was a failure because [the pharmaceutical company] stopped at race. Meaning that [they] should have gone through those subgroups to see what the spectrum of variation was within those groups to see who's going to respond to what, but they stop at th[e] group. . . . So, I s[ee] it as a stall in the movement towards . . . individualized medicine.

So rather than develop and market drugs to specific groups of patients, pharmaceutical companies use the 'race' concept as a category for genetic ancestry to develop medications to ensure they have a bigger marketplace for their medicines. Their drug development and market strategy, however, implicitly implies biological differences underlie socially constructed 'race.' It assumes that an individual's disease susceptibility and severity results from genetic differences found in their specific socially defined racial group. As discussed in chapter 3, genetic ancestry (a synonym for bio-'race') is used in admixture-based genomics to evaluate how gene variation between different socially defined racial groups increases disease susceptibility. In admixture studies these gene variations, which are correlated with geographical ancestry, are used to provide estimates of an individual's genetic ancestry. Because individuals are not genetically ancestrally 'pure,' admixture studies contend that diseases and other phenotypes are outcomes of individuals' mixed genetic ancestry or racial group genetic composition. For example, admixture-based studies of African Americans suggest that genetic variation related to African ancestry might affect both their asthmatic susceptibility and severity based on lung function.[26] These studies claim that the percentage of African ancestry is predictive of lower level lung function in self-identified African Americans compared with that of self-identified 'whites' of the same age, height, and sex.[27] Such research conflates studying the genetic differences with the genetics of racial difference in ways that suggest there are significant biological variations between socially defined groups that require pharmaceutical companies to create specific 'race'-

based drugs for treating aliments for different socially defined groups. Put differently, pharmaceutical companies in the context of a racialized for-profit culture tend to interpret admixture studies relating to disease susceptibility using the genetics of racial difference logic to create personalized medicines. Though some interviewees were well aware their work could be used to market genetic differences as biological racial difference, they did not refrain from using racial grouping as a baseline for differentiating individuals' genomes as evident in this exchange:

> Interviewer: So, from your perspective this kind of race-based practice they're just creating a drug that says, "Okay. Well since these people are from this particular population then they all must be treated by this one particular drug, because they have that in common." Do you think that's a drawback or benefit?
> Interviewee: Well that's a benefit because it can be very targeted, the medication. Basically you are not over medicating some and at the same time you are not getting a useless medication. And to other people you are giving them medication which is very specifically targeted.

But the targeting, which is based on a rather large racial grouping, is far from a precise individual genotypic result given that socially defined 'races' are used, as I argued above, to infer variable traits that characterize the individual's genetic composition. Obviously the targeting ends at the racial group, rather than individual level, as the respondent mentioned at the start of this section.

Given that personal genomic-based medicine is grounded in the idea that human biology is genetics of the particular traits that are passed between generations within a particular socially defined group, interviewees believed that a person is *probably/likely/maybe* genetically a certain way because *most/many/some* individuals belonging to his/her group are that way.[28] Though this logic was used continuously by interviewees, many, like this individual found it difficult to acknowledge that their genomic reasoning used racial genetic similarity as a basis for constructing personal genomic-based medicine treatments:

> Interviewee: . . . I really like personal genomic-based medicine. I mean . . .
> Interviewer: But it's applied to the whole group, it's not just an individual thing.
> Interviewee: Right . . . I do think it good that we . . . I do think it'll be important to use the information that we have about the variants that are out there, and maybe it would associate with certain populations like Jewish populations, or African American populations.
> Interviewer: Yeah?
> Interviewee: I mean, I think it's important that we recognize that there are some sequence differences . . . that can be very useful for designer drugs. I guess I still think calling it race-based medicine is . . . I don't know. I feel uncomfortable with it.

> Interviewer: Problematic?
> Interviewee: It's based . . . It's based on this sequence of the, you know, it's genomic-based medicine.
> Interviewer: Well, let me ask you this. . . . So, there's this tremendous aversion to . . . —I'm getting this from listening to you talk—to even mentioning this idea of race in any kind of research. . . . What's your angst about it? Why are you so concerned about it?
> Interviewee: I guess because genetically race doesn't exist. I mean, I guess it's not a scientific term, is my angst about it.

Despite the informant's concern, personalized genomic medicine's logic usually begins with a particular socially defined racial group identified as having a higher prevalence for a specific disease relevant to genetic variation. As such, racial categories are used to direct researchers to specific groups to uncover the underlying causes of disease.[29] Though these racial categories are merely unsophisticated estimations of a group's genetic variance they implicitly and/or explicitly operate as biological and/or genetic variation identifiers. Consequently, 'race' as a genetic concept gives rise to "racialized notions of biology" in personalized genomic-based medicine.[30] Within the drug development process, racial categories as genetics provide pharmaceutical companies the 'scientific' legitimacy to market their drugs to a racial audience. Like the earlier interviewee, Jonathan Khan contends that targeting a racial audience is cost effective for pharmaceutical companies because they cannot afford to scan every individual's genetic profile.[31] It is far easier to target a specific racial group than to identify a market of particular individuals who happen to respond to a drug well because of their genetic makeup regardless of 'race.'[32] Personalized genomic-based medical researchers, as the first respondent in this section asserted, can say they are using 'race' to get to individualized genomic-based medicine, but pharmaceutical companies are using biology as a surrogate to utilize 'race' in order to bulk market their drugs.[33] In this way, 'good' science of individualized genomics medicine helps to sustain biological notions of 'race' within the discipline and pharmaceutical research and development.

Though we know there is no genetic basis that corresponds with any particular socially defined racial groups, there are genetic characteristics associated with certain populations, but none of these are exclusive, nor correspond uniquely with any one group that might fit a racial category.[34] These genetic features are merely geographical/environment adaptations that are often expressed differently within so-called 'races' but not between them.[35] Because the 'race' concept is a ubiquitous and powerful cultural signifier, genomicists and drug companies tend to superimpose it on social groups' genetic adaptations in ways that suggest racial genetic differences in disease risk and drug metabolism are real. Although biological 'race' does not exist, white supremacy and its associated ideologies of 'race' and whiteness do, and they are closely linked to personalized genomic-based medicine's commercialism. According to Fujimura and her collaborators, the "commercialism of GWAS's results creates a new 'risk genomics' grounded in applying the genetics of populations [read: 'race'] to the genetics of the individual."[36] This commercialized racial character of personalized genomic medicine emerges from the 'good' science of GWAS, which generates single nucleotide polymorphisms (SNPs) of interest that are assumed to be correlated with disease susceptibility. Despite the fact that GWAS has yet to identify the genetic sources of any specific common disease the commercialization the SNPs it finds is expanding. Yet, 'personal genome services' are marketing risk estimates based on uncharacterized SNPs, before geneticists understand if and how these genetic variances relate to disease causation.[37] Hence, Fujimura and her colleagues correctly ask: what exactly are personal genome service and pharmaceutical companies selling given that the relationship between common diseases and genetic variants is still being constituted in the biomedical research context?[38] It seems they are selling 'race'-specific genomics, not personalized genotypic services and treatments tailored to an individual unique genetic code. Personalized genomic companies' marketing schemes promote the idea that 'race' and socially defined racial groups' health are exclusively biological in order to increase the market for their product.

Because the 'race'-specific genomic market for a drug is small and these drugs are not really racially specific, pharmaceutical corporations work to expand the market for their raced products by the promoting 'off-label' use for them to the general population regardless of the consumer's 'race.'[39] This practice involves medical doctors prescribing drugs for ailments not formally specified by the government approving agency. Such prescribing is usually based on clinical studies conducted after the initial drug approval that show its effectiveness for other health problems.[40] In the United States for example, Khan contends that though the government approving agency for drugs, the Food and Drug Administration (FDA), "technically prohibits pharmaceutical corporations from explicitly promoting off-label uses, it does not regulate how doctors prescribe or talk about the possible uses of particular drugs."[41] Pharmaceutical corporations exploit this opening to off-label market by funding "educational seminars" for medical practitioners where they tout research indicating new uses for their drug and solicit doctors to sign on as authors of the studies for sizable "consulting fees."[42] In this way, genomicists and pharmaceutical corporations simultaneously racialize and deracialize personalized genomic medicine as a means for improving the health of general public while highlighting the centrality of 'race' as genetics for an individual's "fitness."

In the context of health services, the fiction that certain biological attributes correlate with 'race' seems like a useful concept, particularly when it is used as a way to identify socially defined racial groups who might benefit from a medical or social intervention. But, as Robert Pollack argues, this apparent usefulness offers up next to nil in understanding common diseases' link to genetic variants given the complexity of human history and the diversity of our genome.[43] Although it is 'bad' genomic science to infer genetic risk for disease by an individual's racial identity, many pharmacogenomicists and pharmaceutical corporations persist in asserting that variation in genes found in one racialized group are susceptible to a particular aliment and not the other, signifying a disease-linked gene.[44] This approach is sustained by both profit logic and racialized genomic methodologies and thinking that largely goes unrecognized by informants, like this individual, who insists that racial ideology does not factor into their understanding of a person's genetic risk for disease:

> You know [race is] ridiculous, offensive, and not scientifically sound. I guess I'm not harboring any secret desire . . . [I]t's not like [I'm] hiding something about not talking about race scientifically. . . . I think talking about genetic differences associated with certain populations can lead to differences in disease states or response to drugs is good. But I wouldn't put a label of race on it.

Though respondents maintained they did not racially label the populations under study for genetic related disease risk and drug response, their discourse indicates that an individual's racial affiliation is a valid means for inferring their genetic composition.[45] Ignoring or not talking about the 'race' concept in personalized genomic-based medical research in effect ensures racial ideology's naturalization within the field. By assuming disease susceptibility variation is linked to racial genetic uniqueness, genomicists and corporations lend legitimacy to the idea that a single genetic difference found in one socially defined racial group and not in other indicates 'race' is a biological phenomenon. This skillful genomic deception is accomplished effortlessly by researchers steering clear of racial ideological discourse through their use of racially innocuous *health improvement* language centered in diagnosing and curing disease at the genetic level. But this discourse is not as unassailable as it seems, according to medical historian Nathaniel Comfort who argues that genomic medicine is closely linked to twenty-first century eugenic science in that modern genomic medicine shares eugenics' basic goal to eliminate all *dangerous* hereditary tendencies.[46] 'Dangerous' in genomic medicine however does not manifest as 'bad' arbitrary diagnoses or draconian policies (e.g., sterilization laws) directed at marginalized socially defined racial and class groups but as 'good' science using targeted 'preventive' measures like genetic screening, selective abortion, and sterilization that conflate the avoidance of disease with the prevention of genetically imperfect racially defined social group members' lives.[47] Though gene sequencing, targeting, and modification technologies are starting to emerge in genomics in a manner that may undercut the need to end lives through gene editing, the form this new application takes is contingent on the dictates and strictures of its particular social and political milieu.

Personalized genomic medicine as 'good medical science' provides genomicists with a means for sustaining their bioracial focus through their use of socially constructed 'race' as a crude marker of an individual's genetic makeup. One has to question why genomicists believe they can use 'social race' as a surrogate for genetic variation given the fact that genomics exist within a racist culture that translates beliefs into racist action. Simply concealing the 'race' concept in genetic discourse and methodologies does not eliminate racial ideology from one's conscious and subconscious since the racial ideology and practices evident in genomics are merely extensions of the racial bigotry and systemic racism permeating society. It is impossible to understand how genomics conceptualizes itself without viewing it as set a practices and beliefs operating across social institutions, public and private.[48] The simple fact is that genomics involves racialized human actors and paradigmatic focuses that come into existence and are sustain through material arrangements and technology. Within this dynamic racist capitalist environment, personalized genomic medicine is constructed as a racial genetic-based medicine presumed to be the most effective way to diagnose and treat illness. The challenge for genomicists is to make racial ideology transparent in the midst of their ordinary genomic variation investigation so they can recognize how not to generate genomic practices and beliefs that produce and sustain systemic racism. In the final analysis, it is not really a matter of separating the 'good' genomic sheep from the 'bad' goats of bias and misuse. Instead, the emphasis should be about understanding genomic variation knowledge's existence within the contexts of systemic racism.[49]

NOTES

1. Paraphrased from Goldcare, *Bad Science: Quacks, Hacks, and Big Pharma Flacks*, iv.
2. Rabinow, ed., *The Foucault Reader: An Introduction to Foucault's Thought*.
3. Paraphrase from Foucault, *Discipline and Punish: The birth of A Prison*.
4. Freedman, "Lies, Damned Lies, and Medical Science."
5. Freedman, "Lies, Damned Lies, and Medical Science."
6. Grinnell, *The Scientific Attitude*.
7. Grinnell, *The Scientific Attitude*.
8. Grinnell, *The Scientific Attitude*, 16.
9. Quantitative genetics is the study of the inheritance of traits that show a continuous distribution of phenotypes in a segregating "population." Traits that are controlled by many genes also exhibit quantitative inheritance as each gene segregates in a Mendelian fashion. Even when there are only a few genes involved, the trait variation will show a continuous distribution due to the results of measurement error and environmental effects.
10. Graves, "Evolutionary versus Racial Medicine."
11. Graves, "Evolutionary versus Racial Medicine."
12. Pickens, *Eugenics and the Progressives*.

13. Abu El-Haj, *The Genetic Reinscription of Race, Annual Review of Anthropology.*
14. Abu El-Haj, *The Genetic Reinscription of Race, Annual Review of Anthropology.*
15. Salber, "Genomics and Personalized Medicine: A Conversation with Peter Donnelly."
16. Offit, "Personalized Medicine: New Genomics, Old Lessons."
17. Offit, "Personalized Medicine: New Genomics, Old Lessons."
18. Lei, "How Will It Change Life? Interview with Dr. Keith Batchelder of Genomic Healthcare Strategies."
19. Joan H. Fujimura et al., "Race and Ancestry: Operationalizing Populations in Human Genetic Variation Studies."
20. Single nucleotide polymorphisms or SNPs (pronounced "snips") are DNA sequence variations that occur when a single nucleotide (A, T, C, or G) is altered. Many SNPs have no known effect on cell function, but some are believed to predispose people to disease or influence their response to a drug.
21. Fujimura et al., "Race and Ancestry: Operationalizing Populations in Human Genetic Variation Studies."
22. Chadarevian, "Genetic Evidence and Interpretation in History."
23. Duster, *Backdoor To Eugenics.*
24. Bourdieu, *The Practice of Reflexive Sociology.*
25. Bourdieu, *The Practice of Reflexive Sociology.*
26. Kumar, et al., "Genetic Ancestry in Lung-Function Predictions."
27. Kumar, et al., "Genetic Ancestry in Lung-Function Predictions.
28. Gannett, "Racism and Human Genome Diversity Research: The Ethical Limits of 'Population Thinking.'"
29. Kahn, "From Disparity to Difference: How Race-Specific Medicines May Undermine Policies to Address Inequalities in Health Care."
30. Kahn, "From Disparity to Difference: How Race-Specific Medicines May Undermine Policies to Address Inequalities in Health Care."
31. Kahn, "From Disparity to Difference: How Race-Specific Medicines May Undermine Policies to Address Inequalities in Health Care."
32. Kahn, "From Disparity to Difference: How Race-Specific Medicines May Undermine Policies to Address Inequalities in Health Care."
33. Kahn, "From Disparity to Difference: How Race-Specific Medicines May Undermine Policies to Address Inequalities in Health Care."
34. Rutherford, "Why Racism Is Not Backed by Science."
35. Rutherford, "Why Racism Is Not Backed by Science."
36. Fujimura et al., "Introduction: Race Genetics and Disease: Questions of Evidence, Matters of Consequence."
37. Fujimura et al., "Introduction: Race Genetics and Disease: Questions of Evidence, Matters of Consequence."
38. Fujimura et al., "Introduction: Race Genetics and Disease: Questions of Evidence, Matters of Consequence."
39. Kahn, *Race in a Bottle: The Story of BiDil and Racialized Medicine in a Post-Genomic Age.*
40. Kahn, *Race in a Bottle: The Story of BiDil and Racialized Medicine in a Post-Genomic Age.*
41. Kahn, *Race in a Bottle: The Story of BiDil and Racialized Medicine in a Post-Genomic Age.*
42. Kahn, *Race in a Bottle: The Story of BiDil and Racialized Medicine in a Post-Genomic Age.*
43. Pollack, "Natural Selections, the Human Genome and the Idea of Race."
44. Pollack, "Natural Selections, the Human Genome and the Idea of Race."
45. Roberts, *Fatal Invention: How Science, Politics, and Big Business Re-create Race in the Twenty-first Century.*
46. Comfort, *The Science of Human Perfection: How Genes Became the Heart of American Medicine.*

47. Comfort, *The Science of Human Perfection: How Genes Became the Heart of American Medicine.*
48. Burchell et al., eds., *The Foucault Effect: Studies in Governmentality, With Two Lectures by and an Interview with Michel Foucault* (Chicago, Illinois: University of Chicago Press, 1991).
49. Haraway, "Situated Knowledges: The Science Question in Feminism and The Privilege of Partial Perspective."

SIX

Reorienting Genomics

Racism and its ideology, like gravity, envelop the world so thoroughly that they operate as an unquestioned truth. Given systemic racism's ubiquity, its ideology is so completely subsumed in genomics that it operates as common sense. For this reason, most genomicists go through the research process without fully grasping racial ideology's impact on how they collect, frame and analyze genetic data. Racial ideology's ordinariness also impedes genomicists' ability to see and understand its operation in the rest of the social world, including education, politics, and other major societal institutions. Unlike gravity, this racialized perceptual and epistemic veil is organized around fictive notions that give rise to racist practices and 'collective consciences' through which genomicists accept racial structure—a network of social relations acting simultaneously at the micro and macro levels—as mundane. Plainly put, 'race' is not a biological fact but an ideological product of systemic racism. Racism is not merely an attitude or a state of mind like bigotry: it's an action.[1] Only when racism is recognized as an exertion of power in genomics can genomicists begin to comprehend how they enshrine socially constructed 'race' as a natural category.

Though genetic evidence overwhelming indicates there is no biological basis for 'race,' I found that racially socialized interviewees maintained, like this individual, that correlative racial thinking occurs within genomics because ". . . obviously there are diseases that have higher incidence in one group that could be defined racially as compared to another, in one community as to defined with . . . as compared with another. And so you can make generalizations. But . . . I would view race more as a sort of cue for [genetic disease risk] . . . ." This assumption is largely a construct of deep beliefs among informants that genetic variance found in many but not all members of a socially defined racial group signifies bioracial differences.

Interviewees proposing socially constructed 'race' as an indicator of alleles associated with disease did so without adequately substantiating why such susceptibilities are dependent on 'social race' because they conceptualize DNA as both a metaphor and attribute for 'race.' This conception runs counter to genome science's concern with knowing the cause and 'reason for' why disease risk and a drug's efficacy vary among social groups. A cause is a principle upon which something depends. Science is about conclusions that carry the force of necessity—if one is uncertain of the truth of a conclusion, one does not yet possess science.[2] A few interviewees admitted, like this individual, that when their colleagues do not provide reasons for their genetic predisposition conclusions they are often less disposed to acknowledge or consider

> that many things about us and in our illnesses, our diseases, are a combination of genetic predispositions we inherited and external factors like diet, lifestyle and environment. We don't understand that very well at the moment. . . . Genetics is part of the story, but not all of the story. . . . As we amass more data and we getter better data on individuals, on their genomic information, on the sorts of diets they have, exercise, and so forth, we will be able to untangle that better.[3]

Though informants were aware that genetics plays only a partial role in disease manifestation, the false notion that bio-'race' is the cause for variation in illness susceptibility remains salient. This is so in part because many interviewees believed that they can statistically make a causal claim that, for example, "black people get this disease, or that black people have more of a certain blood factor than others, with a certain statistical frequency."[4] However, as we learned in chapter 1, it is impossible to "derive a causal explanation from a statistical frequency."[5] To understand why this is the case, let's work through this simple thought analysis on the 'social race'—disease nexus. 'Race' is the independent variable since it is socially contingent thus easily changeable, and disease is the dependent variable. In this superficial observation one finds that there is an 80 percent chance that a certain disease will manifest in people who are socially categorized as black, and that this relationship is statistically significant. But how do genomicists know that being black increases disease risk? Because socially defined 'race' is highly correlated statistically and genomics lack the analytical tools, genomicists cannot explain why there is statistical significance between a specific disease and 'social race.' Statistical analyses alone cannot clarify the relation between socially defined 'race' and disease. Disease causation is the product of variables outside of the statistical data analysis, thus correlative findings are merely suggestive patterns in variation. Conclusions about causation are made outside of the statistical method, given that there are many more nongenetic factors are involved in disease risk. Like all sciences, genomics does not provide conclusive answers, and what little it knows about the human genome and health suggests they are contextually contingent. Hence, the probability that a particular genetic sequence leads to disease manifestation is context dependent as are the statistical methods used to calculate this risk. It is within this setting that interviewees deploy commonsense racial reasoning to argue that disease susceptibility represents racial genetic variance while simultaneously claiming that disease risk variations are also influenced by nongenetic factors. Since the latter is only now starting to gain traction in genomics as a valid and promising avenue for discovering disease etiology, most researchers still center their investigation in establishing 'race' as a biologically meaningful reflection of inherited genetic susceptibility to disease. Despite the empirical weakness of this connection, many interviewees took the existence of 'race' for granted when starting and ending their biomedical research assuming 'race' as a biological entity. This all suggests that researchers "know, and simultaneously don't know," that their claims about 'race' and disease relatedness are grounded in perfunctory reasoning.[6] Such faulty thinking is induced largely by the general racial ideological bent in genomic variation research that takes the form of 'race-based' genetic language, which ascribes racial variation in disease susceptibility to fixed biological traits. This genetic contention derives in part from zoologist Carolus Linnaeus's eighteenth century arbitrary biological classification of humans into 'races' according to shared physical characteristics.[7]

However, the fact that genomic research has not found sharp genetic differentiation among geographical populations did not stop informants from arguing that the genetic similarities of individuals correlate with 'ancestry'—another 'race-based' genetic term. Interviewees' racialized dispositions and assumptions inclined them to misinterpret human displays of geographic, genetic, and physical variation as evidence for the existence of biological 'races.' In addition, their lack of training in evolutionary biology and their incomplete understanding of how evolutionary thinking is shaped by environment played a significant role in generating inaccurate racial explanations about genetic variation among human groups.[8] This knowledge deficit led many respondents to either ignore or downplay the significance of nongenetic factors on human genetic variance. According to this informant, this de-emphasis means that though

> [w]e have good ways now of studying genetics, which we never had ten years ago. We don't have good ways of looking at the environment, classifying the environment, measuring things. Especially, measuring things. . . . You know, the genome is constant . . . and the environment changes rapidly, environment as it affects your biology and my biology and that science is very poorly developed.

Given this analytical deficiency, most interviewees directed their energy into genetic causality when explaining differences in disease susceptibility. This approach led respondents to incorrectly assume that finding a gene that is statistically overrepresented in a socially defined racial group is associated with disease susceptibility, thereby ignoring the fact that the presence of such genes does not really explain the phenotypic characteristic observed. [9] Consequently, though nearly all the respondents rejected the idea that biological differences belie racial distinctions, most continued to categorize genetic variance racially using racial synonyms like admixture, geography and ancestry. These racial genetic euphemisms enabled interviewees to implicitly label the groups they study racially to suggest genetic variation's link to racial identity. Similar to Kaufman's and Cooper's findings over a decade ago, my analysis reveals that "although simplistic biological understandings of race . . . associated with the eugenics movement" are no longer apparent, "the far more subtle presumption that racial . . . distinctions nonetheless capture 'some' meaningful biological differences is alive and flourishing." [10]

Throughout my analysis it is evident that racial thinking about genomics has important implications for the way genomic research is structured and how genome data is collected and evaluated. However, attempts to frame human genetic variance as distinct racial differences encounter problems when confronted with the *principle of discordance*, revealing that individuals in the same socially defined racial group are often not susceptible to similar diseases. That is, reliable prediction of disease predisposition is not possible by the identification of an individual's socially defined 'race.' If this were true it would be possible to develop 'race-specific' treatments. Despite this reality, most informants continue to search for specific disease-related genes presumably shared by socially defined 'races.' Their preoccupation makes no sense because though there is geographically based genetic variance among human groups, there are no 'race-specific' genes. Given this, a select few interviewees like this individual, admitted that genomics' obsession with establishing the existence of biological 'races' may reflect a desire to validate racial hierarchy.

> But the question is why do people try to use [human genetic difference] as a justification for some kind of judgment . . . and that's where the problem is 'cause they intersect a completely non-scientific kind of visceral. . . . I don't know why people do this to one another. I guess it's this native view that you make yourself seem [better] by cutting off other people's heads or something. But inherently we're all . . . , it turns out we know from the HapMap data, the thousand genome data that every one of us has about 60 serious disease genes that could potentially cause about 60 serious conditions. And so there's no such thing as a perfect[/superior] human being. There never has been, and there certainly won't be from this point forward.

Nevertheless, the idea that genetically based differences in disease susceptibility vary by racial grouping persists in human genomic variation research. Why? According to my analysis genomicists hold unacknowledged racialized imaginations that they internalize as guiding truths.[11] Though these patently false racial sensibilities encounter extensive evidence negating their veracity, they remain firmly embedded in genomics. Exactly why this is possible in a scientific inquiry can be gleaned from the insights of sociologist Emile Durkheim's theorizing on praxis and beliefs.[12] Durkheim posited a relationship between praxis and adherence to social order, in this case the racial order. Praxis, organized around 'race,' is made real through the "day-to-day relationships of life" where people racially construe their spaces and bodies (biology).[13] This reflects Durkheim's theorizing regarding how social phenomena become *social facts*. The real physical 'stuff' of 'race' in the form of racist practices matters for how it works to affirm a racial collective conscience that includes genomicists and their compliance with the racial thought and the solidarity it generates. The collective racial experience generated by racism is so powerful that it provides genomicists, and people in general, with a profound sense of connectedness to each other and a deep *white racial frame* of mind that directs how they see themselves and reality. As such, white supremacy usually operates within genomics as an obvious and taken-for-granted reality.

Of course, all people do not experience systemic racism in the same compelling way, as evident among interviewees whose understanding of systemic racism's impact on them and their research varied widely. Respondents' exposure to everyday racism as a ubiquitous force in society marks their varied awareness of white supremacy as a product of this collectively generated system which teaches them to comprehend 'race' as something other than a political ideology emerging from systemic power relations working to sustain racial order. Hence, racism praxis in genomics is best understood as anything genomicists *do* that reminds them of and *renews their belief* in 'race' as a vital and valid means for making sense of the human genome. The collective genomic act of making sense of human genomic variation by racial groupings conveys powerful meanings that recreate and affirm the racial order as ordinary for both genomicists and the public. For 'race' is not a biological category but an ideology constituted through racialized/racist genomic practices. It is only possible for genomicists to begin to comprehend how systemic racist practices and assumptions in genomics produce general beliefs in 'race' when they invert their starting point from 'race' to racism.[14] Doing so will reveal to them the ways in which racialized culture and genomics coproduce one another.

'RACE'-EVASIVE GENOMICS

Although most respondents prided themselves on the 'objectivity' of their research, they spent little time explaining the basis for their confidence, preferring instead to do no more than negate criticisms of those pointing out the racial biases inherent in genomics. Contemporary genomics is a product of a racist eugenic past and current racialized environment; therefore, it is impossible to uncouple genomic knowledge production from racial oppression no matter how strongly genomicists assert their science transcends racism. Their attempts to detach and distance both themselves and genomics from their racist past and operation within the current racialized environment led informants to repeat the mistaken assumptions and practices of their early twentieth century predecessors. Their willful ignorance or refusal to address racial ideology and racist practices in genomics underscored how many interviewees interiorize the logic of white supremacy in ways that ensure their agency as reproducers of systemic racism. As my analysis demonstrates, rather than being immune from racism, genomics is a product of socialized interpreters and decoders whose false detachment and distancing reproduces and reinforces racial ideology and practices that sustain systemic racism without them acknowledging or admitting collusion. Such maneuvering is a manifestation of an induced, socialized, reflexive, collective effort to produce and re-create systemic racism. Informants' refusal to examine or their ignorance of the interplay between racial ideology and genomics led them to construct racially inflected methodologies and analyses of genetic data. As I detailed in my analysis, informants being 'unaware' is a 'race'-evasive tactic deployed to divert attention away from their complicity in white supremacy. It is therefore incumbent on scholars to view racially socialized and aversive genomicists as consistent agents in the reproduction of racialized culture, which equips them with the perceptual and linguistic means to comprehend human genome variation. This was readily apparent among interviewees who sought to dismiss racial ideology's sway in genomics using 'race-based' genetically coded language that confirmed 'race' in the name of avoiding it.

The contention that systemic racism is no longer a problem in genomics mirrors that of color-blind ideology adherents in society at large who maintain that racism is no longer a societal concern. This perspective is generally accepted among genomicists, though no one has yet to empirically substantiate why racism is merely a relic of genomics' past. Systemic racism endures in genomics, in part due to genomicists' belief that science can operate independent of social systems and history. This is humanly impossible. Researchers operate in a racialized cultural context that informs more than their individual actions—it predates them, it is deep-seated, it is institutionally organized and pervasive. Such cultural ubiquity enables racialized genomicists to configure genomic constructs in ways that obscure simple, verifiable facts about genomic variation and disease susceptibility.

Given that the making of group disease risk differences into 'race' is a social act, genomicists' understanding of people's genetic relatedness is constantly remade and transformed in quite often confusing and unsettling racialized ways. Human disease susceptibility, modified as it is in different social and cultural contexts, cannot simply be plugged into a genomic risk algorithm using a single measure as though it can be explained simply by a few genetic factors. Rather a much broader conception of disease susceptibility needs to be formulated and incorporated into genomic research designs. Predisposition to disease is not solely an outcome of a person's genetic composition, but also an individual's interaction within racialized social and cultural environments. In the words of this panelist at the 2006 national conference on genomics, race, and health disparities in San Francisco, California, "racism is real," thus racial attitudes and discrimination affect the different ways genes are activated through such physical factors as stress hormones.[15] Therefore, genomicists need to venture beyond their laboratories to acquire a more realistic understanding of how societies impact genetic functions.[16] There are of course other reasons for genomics' narrow focus on genes as the cause of variation in disease among political racial groupings, but these rationales also skirt examining how meanings about genome variance that do not inhere in the genes themselves are developed through interaction in racialized social contexts. I use aspects of systemic racism theory, social constructionism and its standpoint, and coproduction variants as the conceptual frame to explain how genomicists as individuals and a group contribute to producing and perceiving genomic reality and knowledge. For genes and disease are not simply present in human biology waiting to be discovered by genomicists but created by them acting in and toward reality. In a word, genomicists' understandings of genes are not purely products of genomic reasoning. As the book reveals, the racialized cultural landscape in which interviewees are situated usually eluded them because as anthropologist Ralph Linton once observed, "The last thing a fish would notice is water."[17]

CENTERING RACIAL IDEOLOGY

My study suggests that centering the racialized experiences of interviewees is pivotal to lessening racial thinking in genomics. That acknowledgment that their racialized context plays a role in the production of genomic knowledge is important considering most human genome research is undertaken in response to problems that are racially defined and framed. For example, how do genomics approach the problem of obesity as a significant health concern within human groups? When the solution is to require researchers to adhere to the National Institutes of Health *Policy and Guidelines on the Inclusion of Women and Minorities as Subjects in Clinical Research* outlining the proposed racial composition of their study's population as defined by the Office of Management and Budget (OMB) racial categories, genomics investigations frame the obesity problem as one rooted in genetic differences among socially defined racial groups, downplaying sociopolitical factors, such as racism in inducing obesity. Though understanding socially defined racial group differences in disease susceptibility is important, it is not useful to stop here given gene expression is contingent on chance, environment, and a whole host of other intervening factors. In the case of obesity, genomic evidence suggests that no more than 1 percent of human obesity can be accounted for by genetic variation.[18] Given the rapid increase in obesity after the 1980s, the possibility that this health problem is the result of genetic composition is almost surely zero. A more reasonable explanation is that at about that time people started consuming more calories and continue to do so. In the words of this genetic nutritionist:

> The obesity epidemic we are facing today unfolded over the past few decades and can clearly not be explained by changes in the frequency of risk alleles. It is more likely due to a changing social and physical environment that encourages consumption and discourages expenditure of energy, behaviors that are poorly compatible with the genome that we have inherited.[19]

Nevertheless, most genomic approaches to obesity locate the problem in the individual's biology, rather than in socioeconomic factors contributing to being overweight. This approach to obesity unavoidably leads to the treatment of obesity as a genetic predisposition that focuses on solutions requiring the alteration of the individual rather than the unhealthy diets pushed by the food industry. For instance, the solution offered by Coca-Cola, the largest producer of sugary beverages, is to "get more exercise and worry less about cutting calories."[20] The company allies with influential scientists to convince a weight-conscious public through articles in medical journals, papers delivered at conferences, and social media that they should not be overly fixated on what they eat and drink. To help spread its message, Coca-Cola provides its researcher allies with financial and logistical support to hype the importance of exercise, not diet, in combating obesity. In this way, Coca-Cola uses the imprimatur of science to contend there is "virtually no compelling evidence" that their sugary drinks contribute to the increase in obesity.[21]

This presentation of genes as destiny racializes health problems to the point that the nongenetic factors are largely ignored in explaining the high rates of certain diseases among socially defined racial groups. Because genomics has a penchant for treating socially defined racial groups' health as an individual biological problem rather than as a manifestation of complex racialized sociopolitical and economic factors, the post-genomic era remains squarely centered in genetic claims about 'race' as biological entity. Systemic racism theory, however, suggests genomics should move away from bio-'race' contentions toward conceding the limitation of racial genetic claims regarding disease susceptibility and treatment variance. This change in focus can potentially lead genomicists to be more circumspect about their socially conditioned assertions that 'race'-based genetic inheritance is the source of health differences. Similar to Ann Morning's research, my investigation found that how genomicists conceptualize and interpret human genetic variation knowledge is shaped by their racial assumptions about whose DNA to sample and which types of genetic data to analyze, as well as suppositions about how different populations are related to each other and decisions about which statistical technique to deploy.[22] This distorted conceptualization of genomic variation has far-reaching implications. Beliefs about the biological basis of 'race' affect practices ranging from medical diagnosis and treatment to forensic criminology and genealogy.[23] Some scholars argue that approaching 'race' as biology is necessary because it directs much needed medical resources to underserved racial groups' communities, while others say "it is . . . a mistake to uncritically accept old racial classifications when we study medical treatments. The task is to determine how the social meaning of race can affect biological outcomes."[24]

The best way to begin to remove the fallacious claim that humans are genetically characterized by 'social race' is to carefully consider, as this study does, how racism and racial ideology is conferred as ordinary in genomic practices and thought. This requires genomicists to acknowledge the social character of knowledge production in order to account for racism's influence in genomics. Genomics should be reflexive not because it can be, but because reflexivity aids in clarifying how the human genome works. Accounting for the racialized social conditions of their own knowledge production can quite possibly enable genomicists to engage the distortive effects of these environments.[25] This is not meant to imply that genomics can transcend the racialized context in which it is performed, because it most certainly cannot. But genomicists who are generally unaware of the social conditioning of their knowledge claims can become more aware of it in a reflexive way that can solidify, among genomicists, the importance of complex nongenetic factors in human biology at multiples levels of genomic analysis. Though it is important for genomicists to value reflexivity in order to confront the racial biases in their research, self-awareness is not enough. Problematizing racial thinking in genomics can only be truly accomplished when it is embodied in the research process so that it is practiced as a reflex.[26] With this approach, 'race' as cultural construct becomes the starting point in human genome variation research rather than an end point for evading racism's impact in genomics. To say that 'race' is a cultural construct does not mean it does not exist; social inventions have an objective reality despite their reliance on human thought.[27] In other words, while 'race' is not genetic, racism can certainly affect our biology. Recent research clearly demonstrates that racism can impact the ways our bodies and immune systems develop.[28] For example, researchers recently discovered that African American men experiencing negative effects from racial discrimination have shorter telomeres, the DNA sequences at the ends of chromosomes that are considered biomarkers for biological aging.[29] Over the course of a lifetime telomeres shorten, and previous research links shorter telomeres with an increased likelihood of premature death.[30] Shorter telomeres are also associated with life-shortening diseases such as diabetes, stroke, heart disease, and dementia.[31] This means that though 'race' is not biological, racism can shape our genetics.

The specific challenge in genomics is to unmask the racial ideology encoded in genomics so that it moves forward with its effort to understand the complex biosocial correlates of health. Genomics' response to the challenge must address the two major ways 'race' becomes molecular: the incorporation of systemic racism in the genetics of racialized groups and individuals, and the internalization of racist ideology and practices reinforcing racialized understanding of the human genome. This is best accomplished by genomics abandoning all pretenses that its knowledge production is immune from the racist sociopolitical context in which it is situated. Doing so will enable genomic researchers to transition quickly to the more complex biocultural view of the human genome which methodologically reveals that 'race' is merely a human creation perverting our understandings about biological processes.

NOTES

1. Fields and Fields, "How Race is Conjured: The Fiction of Race Hides the Real Source of Racism and Inequity in America."
2. McManaman, "Why It Is Impossible to Deny the Principle of Causality on the Basis of Quantum Mechanics."
3. Salber, "Genomics and Personalized Medicine: A Conversation with Oxford's Peter Donnelly."
4. Fields and Fields, "How Race is Conjured: The Fiction of Race Hides the Real Source of Racism and Inequity in America."
5. Fields and Fields, "How Race is Conjured: The Fiction of Race Hides the Real Source of Racism and Inequity in America."
6. Fields and Fields, "How Race is Conjured: The Fiction of Race Hides the Real Source of Racism and Inequity in America."
7. Linnaeus, *Systema naturae per regna tria naturae : secundum classes, ordines, genera, species, cum characteribus, differentiis, synonymis, locis.*
8. Trevathan et al., *Evolutionary Medicine and Health.*
9. Graves, "Evolutionary versus Racial Medicine."
10. Kaufman and Cooper, "Commentary: Considerations for the Use of Racial and Ethnic Classification in Etiologic Research."
11. Althusser, *For Marx.*

12. Emilé Durkheim, *The Elementary Forms of the Religious Life.*
13. Durkheim, *The Elementary Forms of the Religious Life.*
14. Fields and Fields, *Racecraft: The Soul of Inequality in American Life.*
15. Hall, "Genomes Trump Race at Health Conference."
16. Hall, "Genomes Trump Race at Health Conference."
17. Conrad, "The Social Construction of Illness: Key Insights and Policy Implications."
18. Li et al., "Cumulative Effects and Predictive Value of Common Obesity-Susceptibility Variants identified by Genome-Wide Association Studies."
19. Bouchard, "Defining the Genetic Architecture of the Predisposition to Obesity."
20. O'Connor, "Coca-Cola Funds Scientists Who Shift Blame for Obesity Away From Bad Diets."
21. O'Connor, "Coca-Cola Funds Scientists Who Shift Blame for Obesity Away From Bad Diets."
22. Mornings, "Does Genomics Challenge the Social Construction of Race?"
23. Mornings, "Does Genomics Challenge the Social Construction of Race?"
24. Duster, "Buried Alive: The Concept of Race in Science."
25. Bourdieu, *Science of Science and Reflexivity.*
26. Bourdieu, *Science of Science and Reflexivity.*
27. Searle, "Social Ontology: Some Basic Principles."
28. Krieger et al., "Racial Discrimination, Psychological Distress, and Self-rated Health among U.S.-born and Foreign-born Black Americans"; Nancy Krieger et al., "Racial Discrimination and Cardiovascular Disease Risk: My Body My Story Study of 1005 US-Born Black and White Community Health Center Participants (US)."
29. Chae et al., "Discrimination, Racial Bias, and Telomere Length in African-American Men."
30. Epel et al., "Accelerated Telomere Shortening in Response to Life Stress."
31. Williams, "Race, Socioeconomic Status, and Health: The Added Effects of Racism and Discrimination."

Appendix: Interview Questions

1. What part of the country are you from?
2. Is that where you spent most of your childhood?
3. What was the racial makeup of your community?
4. Were racial and ethnic identities an important to people in your community?
5. How do you racially identify yourself?
6. What prompted your interests in genetics?
7. What area of genetics are you interested and why?
8. What are the social implications of your research?
9. Do your social and cultural experiences shape your research? If so how, if not why?
10. Do you see your social and cultural dispositions as problem when designing and analyzing genetic studies regarding human variation?
11. What is your definition of species?
12. Before you began genetic research, how did you define race?
13. Do you agree there biological races within the species Homo sapiens? Why or Why not?
14. How do you define race or explain it?
15. What do you think your colleagues generally mean when they use the term race?
16. How genetically different are humans from each other?
17. Do scientists in your subfield discuss the racial implications of their research?
18. Do you ever discuss the topic of biological races? In what context do you discuss it?
19. Do you regularly or periodically converse with other scientists about racialized culture influence in your work? If so why? If not why?
20. Do you consider the racial implications of your work? If so why, if not why not?
21. Are you concerned about racializing your genetic data and conclusions?
22. What specifically can your research tell us about biological races?
23. Do you think focusing on genetic differences among human groups is important?

24. What do you think non-concordance within groups tell us about race? [concordance—*measure of the percentage of* SNPs *that are measured as identical; because races are supposedly genetically identical, it follows that any genetic mutation carried by one would also be carried by others in the group*].
25. From your perspective is there a link between race and disease?
26. Do you think racial categories useful in helping you understand how genes work? If so why? If not why?
27. Do you support or oppose race-based research? Why?
28. There has been a lot of discussion recently about race-based medicine. What do you believe are the benefits and/or drawbacks of using race as a means of treating and grouping patients?
29. What do you want people to understand about human genetics?
30. Do you think conventional scientific ideas of race changed over time?
31. Do you think that current genome research will shape how people think about race? Why or why not?
32. Those are all the questions I have for you. Is there anything else that you would like to add?

References

Nadia Abu El-Haj, "The Genetic Reinscription of Race," *Annual Review of Anthropology* 36 (2007): 283–300.

Kathy S. Albain et al., "Racial Disparities in Cancer Survival among Randomized Clinical Trials Patients of the Southwest Oncology Group," *Journal of the National Cancer Institute* 101 (2009): 984–992.

Ammar Al-Chalabi and Laura Almasy, eds., *Genetics of Complex Human Diseases* (Cold Spring Harbor, NY: Cold Spring Harbor Laboratory Press, 2009).

David Altheide, "Ethnographic Content Analysis," *Qualitative Sociology* 10 (1987): 65–77.

Louis Althusser, "Ideology and Ideological State Apparatuses (Notes towards an Investigation)," in *Lenin and Philosophy and Other Essays*, edited by Fredric Jameson, 85–156. New York: Monthly Review Press, 2001.

Louis Althusser, *For Marx* (London: Penguin Press, 1969).

Alzheimer's Association, "Race, Ethnicity and Alzheimer's Disease," *Alzheimer's and Public Health Spotlight*, April 1, 2013, http://www.alz.org/documents_custom/public-health/spotlight-race-ethnicity.pdf

George Annas, "Genism, Racism, and the Prospect of Genetic Genocide," in *The Future of Values: The 21st Century Talk*, edited by Jérôme Bindé, 284–288 (New York: UNESCO, 2004).

Julie M. Aultman, "Eugenomics: Eugenics and Ethics in the 21st Century," *Genomics, Society and Policy* 2 (2006): 28–49.

Maria Baghramian, "Constructed Worlds, Contested Truths," in *The Problem of Relativism in the Sociology of (Scientific) Knowledge*, edited by Richard Schantz and Markus Seidel, 105–130 (Piscataway, NJ: Transaction Books, 2011).

Elizabeth A. Baker et al., "Addressing Social Determinants of Health Inequities: Learning From Doing," *American Journal of Public Health* 95 (2005): 553–555.

James Baldwin, "White Man's Guilt," *Ebony* (August) 1965: 47–48.

Michael Balter, "Brain Man Makes Waves with Claims of Recent Human Evolution," *Science Magazine* 314 (2006): 1871, 1873.

Eleonora Bartoli et al., *Chasing Colorblindness: White Family Racial Socialization*, (Manuscript submitted for publication, 2014).

K. M. Beaver et al., "Genetic risk, Parent-child Relations, and Antisocial Phenotypes in a Sample of African-American Males," *Psychiatry Research* 175 (2010): 160–164.

Kevin Begos, "Benefactor with a Racist Bent: Wealthy Recluse Apparently Liked the Looks and Potential of Bowman Gray's New Medical-genetics Department," *Winston-Salem Journal*, December 11, 2002.

Ruha Benjamin, *People's Science: Bodies and Rights on the Stem Cell Frontier* (Stanford, CA: Stanford University Press, 2013).

Peter L. Berger and Thomas Luckmann, *The Social Construction of Reality* (New York: Anchor Books, 1967).

Steve Biko, *I Write What I Like* (New York: Harper & Row, 1978).

Catherine Bliss, *Race Decoded: The Genomic Fight for Social Justice* (Stanford, CA: Stanford University Press, 2012).

Carl Boggs, *Gramsci's Marxism* (London: Pluto Press, 1976).

Eduardo Bonilla-Silva, *Racism Without Racists: Color-blind Racism and the Persistence of Racial Inequality in the United States* (Lanham, MD: Rowman & Littlefield, 2003).

Alastair Bonnett, "Anti-racism and the Critique of 'White' Identities," *Journal of Ethnic and Migration Studies* 22 (1996): 97–110.
Claude Bouchard, "Defining the Genetic Architecture of the Predisposition to Obesity," *American Journal Clinical Nutrition* 91 (2010): 5–6.
Pierre Bourdieu, *Science of Science and Reflexivity* (Chicago: University of Chicago Press, 2004).
Pierre Bourdieu and Loïc Wacquant, *An Invitation to Reflexive Sociology* (Chicago: University of Chicago Press, 1992).
Pierre Bourdieu, *The Logic of Practice* (Stanford: Stanford University Press, 1990).
Pierre Bourdieu, *Outline of Theory and Practice* (New York: Cambridge University Press, 1977).
John Bova, "On Relativism, Pragmatism and Critical Theory," *Naked Punch* 6 (2006): 107–122.
Allen Buchanan et al., *From Chance to Choice: Genetics and Justice* (New York: Cambridge University Press, 2000).
R. Burgdorf and M. Burdgorf, "The Wicked Witch is Almost Dead: Buck v. Bell and the Sterilization of Handicapped Persons," *Temple Law Quarterly* 50 (1977): 995–1034.
Vivien Burr, *Social Constructionism*, 2nd edition (New York: Routledge, 2003).
Vivien Burr, *Introduction to Social Constructionism* (New York: Routledge, 1995).
Avashalom Caspi et al., "Role of Genotype in the Cycle of Violence in Maltreated Children," *Science* 297 (2002): 851–854.
Noel Cazenave and Darlene A. Maddern, "Defending the White 'Race': White Male Faculty Opposition to a 'White Racism' Course," *Race and Society* 2 (1999): 25–50.
Soraya de Chadarevian, "Genetic Evidence and Interpretation in History," *BioSocieties* 5 (2010): 301–305.
David H. Chae et al., "Discrimination, Racial Bias, and Telomere Length in African-American Men," *The American Journal of Preventive Medicine* 46 (2014): 103–111.
Daniel Chang, "UM researchers identify new gene associated with Alzheimer's in African Americans," *Miami Herald*, April 10, 2013, accessed February 12, 2014, http://www.miamiherald.com/2013/04/09/3332703/um-researchers-identify-new-gene.html.
Roy A. Clouser, *The Myth of Religious Neutrality: An Essay on the Hidden Role of Religious Belief in Theories* (Notre Dame, IN: University of Notre Dame Press, 1991).
J. Kim-Cohen et al., "MAOA, Maltreatment, and Gene-Environment Interaction in Predicting Children's Mental Health: New Evidence and a Meta-analysis," *Molecular Psychiatry* 11 (2006): 903–913.
Collins, F. "What We Do and Don't Know about 'Race', 'Ethnicity', Genetics and Health at the Dawn of the Genome Era," *Nature Genetics* 36 (2004): S13–S15.
Patricia Hill Collins, "Moving Beyond Gender: Intersectionality and Scientific Knowledge," in *Revisioning Gender*, edited by Myra Ferree, Judith Lorber, and Beth Hess, 261–284 (Thousand Oaks, CA: Sage, 1999).
Nathaniel Comfort, "The Eugenic Impulse," *The Chronicle of Higher Education*, March 4, 2013, http://chronicle.com/article/The-Eugenic-Impulse/135612.
Nathaniel Comfort, *The Science of Human Perfection: How Genes Became the Heart of American Medicine* (New Haven, CT: Yale University Press, 2012).
Peter Conrad, "The Social Construction of Illness: Key Insights and Policy Implications," *Journal of Health and Social Behavior* 51S (2010): S67–S79.
Richard S. Cooper et al., "An International Comparative Study of Blood Pressure in Populations of European vs. African Descent," *BMC Medicine* 3 (2005): 1–8.
Richard S. Cooper et al., "The Puzzle of Hypertension in African-Americans," *Scientific American* 280 (1999): 56–63.
Francis H. C. Crick, "On Protein Synthesis," *Symposia of the Society for Experimental Biology* 12 (1958): 138–163.

Danielle Deaver, "Forsyth in the Forefront," *Winston-Salem Journal*, December 9, 2002, accessed July 4, 2012, http://www.journalnow.com/news/local/forsyth-in-the-forefront/article_647de82c-8c02-11e2-affa-001a4bcf6878.html.

Gary DeMar, "The Myth of Objectivity," *The American Vision*, December 4, 2006, http://americanvision.org/1360/myth-of-objectivity.

Adrian Desmond and James Moore, *Darwin's Sacred Cause: How a Hatred of Slavery Shaped Darwin's Views on Human Evolution* (New York: Houghton Mifflin Harcourt, 2009).

Paul DiMaggio, "Culture and Cognition," *Annual Review of Sociology* 23 (1997): 263–287.

Theodosius Dobzhansky, *Genetics of the Evolutionary Process* (New York: Columbia University Press, 1970).

W. E. B. Du Bois, *The Souls of Black Folk* (Mineola, NY: Dover, [1903] 1994).

W. E. B. Du Bois, *Dusk of Dawn* (New York: Harcourt Brace Jovanovich, 1940).

Will Dunham, "Genetic Links to Prostate Cancer," *Reuters*, April 2, 2007, accessed August 4, 2011, http://www.reuters.com/article/2007/04/02/idUSN303173732007 0402.

Thomas Dunn and Theodosius Dobzhansky, *Heredity, Race, and Society* (New York: Pelican, 1946).

Emilé Durkheim, *The Elementary Forms of the Religious Life* (New York: Oxford University Press, [1915] 2001).

Troy Duster, "Buried Alive: The Concept of Race in Science," *Chronicle of Higher Education*, September 14, 2001, http://chronicle.com/article/Buried-Alive-the-Concept-of/20283.

Troy Duster, *Backdoor to Eugenics* (New York: Routledge, 2003).

Jennifer L. Eberhardt et al., "Seeing Black: Race, Crime, and Visual Processing," *Journal of Personality and Social Psychology* 87 (2004): 876–893.

Catherine Elton, "Why Racial Profiling Persists in Medical Research," *Time Magazine*, August 22, 2009, accessed June 2015, http://content.time.com/time/health/article/0,8599,1916755,00.html.

Elissa S. Epel et al., "Accelerated Telomere Shortening in Response to Life Stress," *Proceeding of the National Academy of Sciences of the United States* 101 (2004): 17312–17315.

Paul E. Farmer et al., "Structural Violence and Clinical Medicine," *PLoS Med 3*, October 24, 2006, http://journals.plos.org/plosmedicine/article?id=10.1371/journal.pmed.0030449.

Joe R. Feagin, *Racist America: Roots, Current Realities and Future Reparations* (New York: Routledge, 2014).

Joe R. Feagin and Sean Elias, "Rethinking Racial Formation Theory: A Systemic-Racism Critique." *Ethnic and Racial Studies* 36 (2013): 931–960.

Joe R. Feagin, *The White Racial Frame: Centuries of Racial Framing and Counter-Framing,* (New York: Routledge, 2010).

Joe R. Feagin, *Systemic Racism: A Theory of Oppression* (New York: Taylor & Francis, 2006).

Margery Fee, "Racializing Narratives: Obesity, Diabetes and the 'Aboriginal' Thrifty Genotype," *Social Science & Medicine* 62 (2006): 2988–2997.

Barbara J. Fields and Karen E. Fields, "How Race is Conjured: The Fiction of Race Hides the Real Source of Racism and Inequity in America," *Jacobin*, June 29, 2015, accessed July 1, 2015, https://www.jacobinmag.com/2015/06/karen-barbara-fields-racecraft-dolezal-racism.

Karen E. Fields and Barbara J. Fields, *Racecraft: The Soul of Inequality in American Life* (London: Verso Books, 2014).

Michel Foucault, *Discipline and Punish: The Birth of a Prison* (London: Penguin, 1991).

Michel Foucault, *Power/Knowledge: Selected Interviews and Other Writings,* (New York: Pantheon, 1980).

Reanne Frank, "What to Make of It? The (Re)emergence of a Biological Conceptualization of Race in Health Disparities Research," *Social Science and Medicine* 64 (2007): 1977–1983.

John Hope Franklin, *The Color Line: Legacy for the Twenty-first Century* (Columbia, MO: University of Missouri Press, 1993).

David Freedman, "Lies, Damned Lies, and Medical Science," *The Atlantic*, November 2010, accessed August 2, 2015, http://www.theatlantic.com/magazine/archive/2010/11/lies-damned-lies-and-medical-science/308269.

Augustin Fuentes, "Is Aggression Genetic," *Salon Magazine*, May 8, 2012, accessed July 22, 2014, http://www.salon.com/2012/05/28/is_aggression_genetic.

Joan H. Fujimura and Ramya Rajagoplan, "Different Differences: The Use of 'Genetic Ancestry' versus Race in Biomedical Human Genetic Research," *Social Studies of Science* 41 (2011): 5–30.

Joan H. Fujimura et al., "Race and Ancestry: Operationalizing Populations in Human Genetic Variation Studies, in *What's the Use of Race? Modern Governance and the Biology of Difference*, edited by Ian Whitmarsh and David S. Jones, 169–183. Cambridge, MA: MIT Press, 2010.

Joan Fujimura et al., "Introduction: Race, Genetics, and Disease: Questions of Evidence, Matters of Consequence," *Social Studies of Science* 38 (2008): 643–656.

Duana Fullwiley, "The 'Contemporary Synthesis': When Political Inclusive Science Relies on Biological Notions of Race," *History of Science and Society* 105 (2014): 803–814.

Duana Fullwiley, "The Biological Construction of Race: 'Admixture' Technology and the New Genetic Medicine," *Social Studies of Science* 38 (2008): 695–735.

Duana Fullwiley, "The Molecularization of Race: Institutionalizing Human Difference in Pharmacogenetics Practices," *Science as Culture* 16 (2007): 1–30.

Duana Fullwiley, "Race and Genetics: Attempts to Define the Relationship," *BioSocieties* 2 (2007): 221–237.

Francis Galton, *Hereditary Genius* (New York: Macmillan, [1869] 2005).

Lisa Gannett, "The Biological Reification of Race," *British Society for the Philosophy of Science* 55 (2004): 323–345.

Lisa Gannett, "Racism and Human Genome Diversity Research: The Ethical Limits of 'Population Thinking,'" *Philosophy of Science* 68 (2001): S479–S492.

Lisa Gannett, "Racism and Human Genome Diversity Research: The Ethical Limits of 'Population Thinking,'" *Philosophy of Science* 68 (2000): S479–S492.

Kenneth L. Garver and Bettylee Garver, "Eugenics: Past, Present, and the Future," *American Journal of Human Genetics* 49 (1991): 1109–1118.

Jane Gitschier, "The Whole Side of It – An Interview with Neil Risch," *PLoS Genetics* 1 (2005): 3–5.

Brooke Gladstone, *The Influencing Machine*, (New York: W. W. Norton, 2011).

Vicki Glaser, "Investigator Profile: An Interview with Keith Cheng, M.D., Ph.D," *Zebrafish* 3 (2006): 1–7.

Ben Goldcare, *Bad Science: Quacks, Hacks, and Big Pharma Flacks* (New York: Faber and Faber, 2008).

Thomas F. Gossett, *Race: The History of an Idea in America* (New York: Oxford University Press, 1997).

Stephen Jay Gould, *The Mismeasure of Man* (New York: W. W. Norton, 1981).

Stephen Jay Gould, *The Lying Stones of Marrakech: Penultimate Reflections in Natural History* (New York: Three Rivers Press, 2000).

Graham Burchell et al., eds., *The Foucault Effect: Studies in Governmentality, With Two Lectures by and an Interview with Michel Foucault* (Chicago: University of Chicago Press, 1991).

Antonio Gramsci, *Selections from the Prison Notebooks of Antonio Gramsci* (New York: International Publishers, 1971).

Joseph L. Graves, "Why the Nonexistence of Biological Races Does Not Mean the Nonexistence of Racism," *American Behavioral Scientist* 59 (2015): 1474.

Joseph L. Graves, "Evolutionary versus Racial Medicine," in *Race and the Genetic Revolution: Science, Myth, and Culture,* edited by Sheldon Krimsky and Kathleen Sloan, 142–170 (New York: Columbia University Press, 2011).

Joseph Graves, *The Race Myth: Why We Pretend Race Exists in America* (New York: Dutton, 2005).

Joseph L. Graves, *The Emperor's New Clothes: Biological Theories of Race at the Millennium* (New Brunswick, NJ: Rutgers University Press, 2001).

Fred Grinnell, *The Scientific Attitude* (New York: Guilford Press, 1992).

Carl T. Hall, "Genomes Trump Race at Health Conference," *San Francisco Chronicle,* August 19, 2006, B1.

Stuart Hall, "Signification, Representation, Ideology: Althusser and the Post-Structualist Debates," *Critical Studies in Mass Communication* 2 (1985): 91–114.

Jennifer A. Hamilton, "Revitalizing Difference in the HapMap: Race and Contemporary Genetic Variation Research," *Journal of Law, Medicine, & Ethics* 36 (2008): 471–477.

Donna Haraway, "Situated Knowledges: The Science Question in Feminism and the Privilege of Partial Perspective," *Feminist Studies* 14 (1988): 574–599.

Sandra Harding, *Whose Science? Whose Knowledge? Thinking from Women's Lives* (Ithaca, NY: Cornell University Press, 1991).

Sandra Harding, "Introduction: Is There a Feminist Method>" in *Feminism and Methodology,* edited by Sandra Harding, 1–14. (Bloomington: Indiana University Press, 1987).

Jenée Desmond-Harris, "The Biggest Lie About Race? That It's Real," *The Root,* July 26, 2011, accessed June 10, 2015, http://www.theroot.com/articles/culture/2011/07/lies_about_race_dorothy_roberts_and_fatal_invention.1.html.

Anthony Q. Hazard, Jr., *Postwar Anti-Racism: The United States, UNESCO, and "Race," 1945–1968* (New York: Palgrave Macmillan, 2012).

Mary Hebert and Doug Turnbull, "Mitochondrial Replacement to Prevent the Transmission of Mitochondrial DNA Disease," *EMBO Reports* 16 (2015): 539–540.

David T. Helm et al., "Prenatal Diagnosis of Down Syndrome: Mother's Reflection on Supports Needed from Diagnosis to Birth," *Mental Retardation* 36 (1998): 55–61.

Jane E. Helms, "Introduction: Review of Racial Identity," in *Black and White Racial Identity: Theory, Research, and Practice,* edited by Janet E. Helms, 3–8 (Westport, CT: Greenwood Press, 1993).

Lawrence A. Hirschfeld, "Children's Developing Conceptions of Race," in *Handbook of Race, Racism, and the Developing Child,* edited by Steven M. Quintana and Clark McKown, 37–54 (Hoboken, NJ: John Wiley & Sons, 2008).

Laura Hix. "Modern Eugenics: Building a Person?," *Helix Magazine,* July 23, 2009, accessed May 11, 2015, http://helix.northwestern.edu/article/modern-eugenics-building-better-person.

Linda M. Hunt and Mary S. Megyesi, "The Ambiguous Meanings of the Racial/Ethnic Categories Routinely used in Human Genome Research," *Social Science of Medicine* 66 (2008): 349–361.

Nina Jablonski, "The Evolution of Skin and Skin Color," *Annual Review of Anthropology* 33 (2004): 585–623.

Fatimah L. Jackson, "Anthropological Measurement: The Mismeasure of African Americans," *American Academy of Political and Social Science* 568 (2000): 154–171.

John P. Jackson, Jr. and Nadine M. Weidman, *Race, Racism, and Science: Social Impact and Interaction* (New Brunswick, NJ: Rutgers University Press, 2006).

Patricia Jacobs et al., "Aggressive Behavior, Mental Subnormality and the XYY Male," *Nature* 2008 (1965): 1351–1352.

Shelia Jasanoff, "A New Climate for Society," *Theory, Culture & Society* 27 (2010): 233–253.

Jewish Currents, "Genetics, Race, and Socially Responsible Science: An Interview with Alan McGowan," *Jewish Currents,* November 2003, accessed June 15, 2015, http://jewishcurrents.org/genetics-race-and-socially-responsible-science-15541.

Fard Johmar, "Interview Subject: Dr. Esteban González Burchard," *Science, Biology, Evolution*, February 2006, http://www.gatago.com/sci/bio/evloution/43559606.html.

Allan G. Johnson, *Privilege, Power and Difference* (Boston: McGraw-Hill, 2001).

Lynn Jorde et al., "The Distribution of Human Genetic Diversity: A Comparison of Mitochondrial, Autosomal, and Y-chromosome Data," *American Journal of Human Genetics* 66 (2000): 979–988.

Jonathan Kahn, "How a Drug Becomes 'Ethnic': Law, Commerce, and the Production of Racial Categories in Medicine," *Yale Journal of Health, Law, and Ethics* 4 (2013): 1–46.

Jonathan Kahn, *Race in a Bottle: The Story of BiDil and Racialized Medicine in a Post-Genomic Age* (New York: Columbia University Press, 2013).

Jonathan Kahn, "From Disparity to Difference: How Race-Specific Medicines May Undermine Policies to Address Inequalities in Health Care," *Southern California Interdisciplinary Law Journal* 15 (2005): 105–130.

Ulrich Kattmann, "Race, Genes, and Racism" (Proceedings of CAIR10, the First Conferences on Applied Interculturality Research, Graz, Austria: April 7–10, 2010).

Phyllis. A. Katz and Jennifer. A. Kofkin, "Race, Gender, and Young Children," in *Developmental Psychopathology: Perspectives on Adjustment, Risk, and Disorder*, edited by Suniya S. Luthar, Jacob. A. Burack, Dante Cicchetti and John R Weisz, 51–74 (New York: Cambridge University Press, 1997).

Jay S. Kaufman, "The Anatomy of a Medical Myth," *Social Science Research Council Web Forum: Is Race Real*, June 7, 2006, http://raceandgenomics.ssrc.org/Kaufman.

Jay S. Kaufman and Richard S. Cooper, "Commentary: Considerations for the Use of Racial and Ethnic Classification in Etiologic Research," *American Journal of Epidemiology* 154 (2001): 291–298.

Daniel J. Kevles, "Eugenics and the Human Genome Project," *Contention: Debates in Society, Culture, and Science* 2 (1993): 21–37.

Daniel J. Kevles, *In the Name of Eugenics: Genetics and the Uses of Human Heredity* (New York: Knopf, 1985).

Razib Khan, "Interview with Bruce Lahn," *Gene Expression*, July 10, 2006, http://www.gnxp
.com/blog/2006/07/10-questions-for-bruce-lahn_10.php.

Razib Khan, "Interview with Luigi Luca Cavalli-Sforza," *Gene Expression*, July 10, 2006, http://www.gnxp.com/blog/2006/07/10-questions-for-luigi-lucacavalli.php.

David King et al., "Eugenics Fear over Gene Modification," *The Guardian*, March 15, 2013, accessed March 16, 2013, http://www.theguardian.com/science/2013/mar/15/eugenics-fear-over-gene-modification.

Gina Kolata, "In Blacks, Alzheimer's Study Finds Same Variant Genes as in Whites," *New York Times*, April 9, 2013, accessed April 20, 2013, http://www.nytimes.com/2013/04/10/health/african-americans-have-higher-risk-of-alzheimers-study-shows.html.

Nancy Krieger et al., "Racial Discrimination and Cardiovascular Disease Risk: My Body My Story Study of 1005 US-Born Black and White Community Health Center Participants (US)," *PLoS One*, October 18, 2013, http://journals.plos.org/plosone/article?id=10.1371/journal.pone.0077174.

Nancy Krieger et al.,"Racial Discrimination, Psychological Distress, and Self-rated Health among U.S.-born and Foreign-born Black Americans," *American Journal of Public Health* 101 (2011): 1704–1713.

Nancy Krieger, "Stormy Weather: Race, Gene Expression, and the Science of Health Disparities," *American Journal of Public Health* 95 (2005): 2155–2160.

Thomas Kuhn, *The Structure of Scientific Revolution* (Chicago: University of Chicago Press, 2012).

Rajesh Kumar, et al., "Genetic Ancestry in Lung-Function Predictions," *New England Journal of Medicine* 363 (2010): 321–330.

Bruno Latour, *Pandora's Hope: Essays on the Reality of Science Studies* (Cambridge, MA: Harvard University Press, 1999).

Harry Laughlin, *Eugenical Sterilization in the United States* (Chicago: Psychopathic Laboratory of the Municipal Court of Chicago, 1922).
Sandra Soo-Jin Lee, "Biobanks of a 'Racial Kind': Mining for Differences in the New Genetics," *Patterns of Prejudice* 40 (2006): 443–460.
Sally Lehrman, "Race, Genes, and Illness," *Boston Globe*, April 19, 2007, accessed February 26, 2013, http://www.boston.com/news/globe/editorial_opinion/oped/articles/2007/04/19/race_genes_and_illness.
Hisen-Hsien Lei, "How Will It Change Life? Interview with Dr. Keith Batchelder of Genomic Healthcare Strategies," *Eye on DNA*, May 1, 2007, http://www.eyeondna.com/2007/05/01/interview-with-dr-keith.
Daniel Lende, "Nicholas Wade and His Determinist Genes," *PLoS Blog*, July 3, 2014, http://blogs.plos.org/neuroanthropology/2014/07/03/nicholas-wade-determinist-genes.
Armand Leroi, "A Family Tree in Every Gene," *New York Times*, March 14, 2005, A21.
Amanda Lewis, "'What Group?' Studying Whites and Whiteness in an Era of 'Colorblindness,'" *Sociological Theory* 22 (2004): 623–646.
Richard C. Lewontin, "The Apportionment of Human Diversity," *Evolutionary Biology* 6 (1972): 381–398.
Richard C. Lewontin, "Confusions about Human Races," *SSRC Web Forum Is Race 'Real,'* June 7, 2006, http://raceandgenomics.ssrc.org/Lewontin.
Richard C. Lewontin, *Biology as Ideology: The Doctrine of DNA* (New York: Harper Perennial, 1992).
Shenguxu Li et al., "Cumulative Effects and Predictive Value of Common Obesity-Susceptibility Variants Identified by Genome-Wide Association Studies," *American Journal of Clinical Nutrition* 91 (2010): 184–190.
Grace Lichtenstein, "Funds Backs Controversial Study of 'Racial Betterment,'" *New York Times*, December 11, 1977.
M. Susan Lindee et al., "Introduction: Anthropology in the Age of Genetics – Practice, Discourse, and Critique," in *Genetic Nature/Culture: Anthropology and Science Beyond the Two-Culture Divide*, edited by M. Susan Lindee, Alan H. Goodman, and Deborah Heath, 1–19 (Berkeley: University of California Press, 2004).
Carolus Linnaeus, *Systema naturae per regna tria naturae : secundum classes, ordines, genera, species, cum characteribus, differentiis, synonymis, locis* (Stockholm: Laurentius Salvius, 1758).
Abby Lippman, "Prenatal Genetic Testing and Geneticization: Mother Matters for All," *Fetal Diagnosis and Therapy* 8 (1993): 175–188.
Abby Lippman, "Prenatal Genetic Testing and Screening: Constructing Needs and Reinforcing Inequities," *American Journal of Law and Medicine* 1 (1991): 15–50.
George Lipsitz, *The Possessive Investment in Whiteness: How White People Profit from Identity Politics* (Philadelphia: Temple University Press, 2006).
George Lipsitz, "The Possessive Investment in Whiteness: Racialized Social Democracy and the 'White' Problem in American Studies," *American Quarterly* 47 (1995): 369–387.
Margaret Lock, "The Tempering of Medical Anthropology: Troubling Natural Categories," *Medical Anthropology Quarterly* 15 (2001): 478–492.
Paul Lombardo. A third generation of imbeciles? *DNA Learning Center Cold Spring Harbor Laboratory*, http://www.dnalc.org/view/15240-A-third-generation-of-imbeciles-Paul-Lombardo.html.
Paul A. Lombardo, "'The American Breed': Nazi Eugenics and the Origins of the Pioneer Fund," *Albany Law Review* 65 (2001–2003) 65: 743–825.
Jeffrey C. Long and Rick A. Kittles, "Human Genetic Diversity and the Nonexistence of Biological Races," *Human Biology* 75 (2003): 449–471.
Salvador Edward Luria, "Human Genome Program," *Science* 246 (1989): 873.
Calum MacKellar, "Ethics and Genetics of Human Behavior," *The Center for Bioethics and Human Dignity*, accessed May 8, 2008, http://www.cbhd.org/resources/genetics/mackellar_2007-04-20_print.htm.

John McDermott, "Lunch with Junot Díaz," *Financial Times,* July 4, 2014, accessed July 10, 2014, http://www.ft.com/intl/cms/s/2/a438f98e-01f4-11e4-bb71-00144feab7de.html#axzz36guL0FS5.

Monica McDermott and Frank L. Samson, "White Racial and Ethnic Identity in the United States," *Annual Review of Sociology* 31 (2005): 245–261.

Sarah McDowell et al., "Systematic Review and Meta-analysis of Ethnic Differences in Risks of Adverse Reactions to Drugs Used in Cardiovascular Medicine," *BMJ Online First* 332 (2006): 1177, accessed March 3, 2011, doi:http://dx.doi.org/10.1136/bmj.38803.528113.55

Peter McLaren, "Whiteness Is . . . The Struggle for Postcolonial Hybridity," in *White Reign: Deploying Whiteness in America,* edited by J. Kincheloe, S. Sheinberg, N. Rodgriuez and R. Channault, 6–76 (New York: St. Martin, 1998).

Den D. McManaman, "Why It Is Impossible to Deny the Principle of Causality on the Basis of Quantum Mechanics," *York Catholic District School Board,* accessed December 13, 2014, http://fmmh.ycdsb.ca/teachers/fmmh_mcmanaman/pages/tok_denying_causality.pdf.

Kenan Malik, "Why Both Sides are Wrong," *Pandaemonium,* March 12, 2012, https://kenanmalik.wordpress.com/2012/03/04/why-both-sides-are-wrong-in-the-race-debate.

Kenan Malik, "Analysis: Colour Coded Prescription," *Kenan Malik Show,* November 17, 2005, http://www.kenanmalik.com/tv/analysis_race%2Bmedicine.html.

Karl Mannheim, *Ideology and Utopia* (New York: Harcourt Brace Jovanovich, 1936).

Elizabeth Martinez, "What is White Supremacy?" Catalyst Project, January 1, 2013, accessed December 30, 2015, http://collectiveliberation.org/wp-content/uploads/2013/01/What_Is_White_Supremacy_Martinez.pdf

Karl Marx, *The Economic and Philosophic Manuscripts of 1844,* trans. Dirk J. Struik (New York: International Publishers, 1964).

Barry Mehler, "The New Eugenics: Academic Racism in the U.S. Today," *Science for the People* 15 (1983): 18–23.

Barry Mehler, "Eliminating the Inferior: American and Nazi Sterilization Programs," *Science for the People* 12 (1987): 14–18.

Pankaj Mehta, "There's a Gene for That," *Jacobin,* June 1, 2014, accessed June 1, 2014, https://www.jacobinmag.com/2014/01/theres-a-gene-for-that.

Frank Miele, "Darwin's Dangerous Disciples: An interview with Richard Dawkins," *Scepsis: Magazine of Science and Social Criticism,* May 2003, accessed May 29, 2015, http://scepsis.ru/eng/articles/id_3.php.

Robert Miles, *Racism* (London: Routledge, 1989).

Charles W. Mills, *The Racial Contract* (Ithaca, NY: Cornell University Press, 1997).

Ashley Montagu, *Man's Most Dangerous Myth: The Fallacy of Race* (New York: Columbia University Press, 1945).

Michael J. Montoya, "Bioethnic Conscription: Genes, Race, and Mexicana/o Ethnicity in Diabetes Research," *Cultural Anthropology* 22 (2007): 94–128.

Richard Moran, "The Search for Born Criminal and the Medical Control of Criminality," in *Deviance and Medicalization: From Badness to Sickness,* edited by Peter Conrad and Joseph Schneider, 215–240 (Philadelphia: Temple University Press, 1992).

Ann Morning, "And You Thought We Had Moved Beyond All That: Biological Race Returns to the Social Sciences," *Ethnic and Racial Studies* 37 (2014): 1676–1685.

Ann Mornings, "Does Genomics Challenge the Social Construction of Race?," *Sociological Theory* 32 (2014): 189–207.

National Academy of Achievement, "Interview: Ernst Mayer: The Darwin of the 20th Century," *National Academy of Achievement,* April 5, 2001, http://www.achievement.org/autodoc/printmember/may1int-1.

National Cancer Institute, "FDA Approval of Gefitinib," National Institutes of Health, January 18, 2011, accessed July 14, 2015, http://www.cancer.gov/cancertopics/treatment/drugs/fda-gefitinib.

Masatoshi Nei and A. K. Roychoudhury, "Genic Variation Within and Between the Three Major Races of Man: Caucasoids, Negroids, and Mongoloids," *American Journal of Human Genetics* 26 (1974): 421–443.

Alondra Nelson, "Bio Science Genetic Genealogy Testing and the Pursuit of African Ancestry," *Social Studies of Science* 38 (2008): 759–783.

Sanne Nikkelen et al., "Media Violence and Children ADHD-Related Behaviors: A Genetic Susceptibility Perspective," *Journal of Communication* 64 (2014): 42–60.

Osagie K. Obasogie, *Blinded by Sight: Seeing Race through the Eyes of the Blind* (Palo Alto, CA: Stanford University Press, 2014).

Anahad O'Connor, "Coca-Cola Funds Scientists Who Shift Blame for Obesity Away From Bad Diets," *New York Times,* August 9, 2015, accessed August 9, 2015, http://well.blogs.nytimes.com/2015/08/09/coca-cola-funds-scientists-who-shift-blame-for-obesity-away-from-bad-diets.

Kenneth Offit, "Personalized Medicine: New Genomics, Old Lessons," *Human Genetics* 230 (2011): 3–14.

Catherine M. O'Leary, "Counteridentification of Counterhegemony? Transforming Feminist Standpoint Theory," *Women and Politics* 18 (1997): 945–972.

Yin C. Paradies et al., "Racialized Genetics and the Study of Complex Diseases," *Perspectives in Biology and Medicine* 50 (2007): 203–227

Erik Parens and Adrienne Asch, "Disability Rights Critique of Prenatal Genetic Testing: Reflections and Recommendations," *Mental Retardation and Developmental Disabilities Research Review* 9 (2003): 40–47.

Suey Park, "Challenging Racism and the Problem with White "Allies": A Conversation with David Leonard" *{Young}ist.org,* December 26, 2013, http://youngist.org/challenging-racism-and-the-problem-with-white-allies/#.VexLI_SF8sA.

Karl Pearson, *The Grammar of Science* (New York: Ulan Press, [1892] 2011).

Donald K. Pickens, *Eugenics and the Progressives* (Nashville, TN: Vanderbilt University Press, 1968).

Robert Pollack, "Natural Selection, the Human Genome, and the Idea of Race," *GeneWatch,* July 27, 2009, accessed February 20, 2015, http://www.councilforresponsiblegenetics.org/GeneWatch/GeneWatchPage.aspx?pageId=199.

Mica Pollock, *Colormute: Race Talk Dilemmas in an American School* (Princeton, NJ: Princeton University Press, 2004).

William B. Provine, "Geneticists and the Biology of Race Crossing," *Science* 182 (1973): 790–796.

Paul Rabinow, "Artificiality and Enlightenment: From Sociobiology to Biosociality," in *Incorporations,* edited by J. Carary and S. Kwiner, 234–252 (New York: Zone Books, 1992).

Paul Rabinow, ed., *The Foucault Reader: An Introduction to Foucault's Thought* (London: Penguin, 1991).

Jenny Reardon, "Finding Oprah's Roots, Losing the World: Beyond the Liberal Anti-Racist Genome" (Unpublished paper presented at Berkeley Workshop on Environmental Politics, October 23, 2009).

Jenny Reardon, *Race to the Finish: Identity and Governance in an Age of Genomics* (Princeton, NJ: Princeton University Press, 2005).

Jenny Reardon, "Decoding Race and Human Difference in a Genomic Age," *A Journal of Feminist Cultural Studies* 15 (2004): 38–65.

Adolph Reed, Jr., "Marx, Race, and Neoliberalism," *New Labor Forum* 22 (2013): 49–57.

Wilhelm Reich, *The Mass Psychology of Fascism* (New York: Simon & Schuster, 1970).

I. M. Reit et al., "Monoamine Oxidase A regulates Antisocial Personality in Whites with No History of Physical Abuse," *Comprehensive Psychiatry* 52 (2011): 188–194.

Christianne Retiz et al., "Variants in the ATP-Binding Cassette Transporter (ABCA7), Apolipoprotein Eε4, and the Risk of the Late-Onset Alzheimer's Disease in African Americans," *Journal of the American Medical Association* 309 (2013): 1483–1492.

Neil Risch et al., "Categorization of Humans in Biomedical Research: Genes, Race, and Disease," *Genome Biology* 3 (2002): 1–12.

Dorothy Roberts, *Fatal Invention: How Science, Politics, and Big Business Re-create Race in the Twenty-first Century* (New York: The New Press, 2011).

John A. Robertson, "Extending Preimplantation Genetic Diagnosis: The Ethical Debate," *Human Reproduction* 18 (2003): 465–471.

Richard Rorty, *Philosophy and the Mirror of Nature* (Princeton, NJ: Princeton University Press, 1979).

Nikolas Rose, "Introduction to the Discussion of Race and Ethnicity in Nature Genetics," *BioSocieties* 1 (2006): 307–311.

Nikolas Rose, "The Politics of Life Itself," *Theory, Culture, & Society* 18 (2001): 1–30.

N. A. Rosenberg, et al., "Genetic Structure of Human Populations," *Science* 298 (2002): 2381–2385.

S. Rosenberg et al, "The Association of DNA Sequence Variation at the MAOA Genetic Locus with Quantitative Behavioral Traits in Normal Males," *Human Genetics* 120 (2006): 447–459.

Charmaine D. M. Royal and Georgia M. Dunston, "Changing the Paradigm from 'Race' to Human Genome Variation," *Nature* 36 (2004): S5–S7.

Adam Rutherford, "Why Racism Is Not Backed by Science," *The Guardian*, March 15, 2015, accessed March 16, 2015, http://www.theguardian.com/science/2015/mar/01/racism-science-human-genomes-darwin.

Dennis M. Rutledge, "Socialization and Racism: The White Experience," in *Impacts of Racism on White Americans*, edited by Benjamin P. Bowser and Raymond G. Hunt, 71–85 (London: Sage, 1981).

Patricia Salber, "Genomics and Personalized Medicine: A Conversation with Peter Donnelly," *The Doctor Weighs In*, February 5, 2015, http://thedoctorweighsin.com/genomics-and-personalized-medicine-peter-donnelly.

Ricardo V. Santos and Marcos C. Maio, "Race, Genomics, Identities and Politics in Brazil," *Critique of Anthropology* 24 (2004): 247–378.

Jean-Paul Sartre, *Critique of Dialectical Reason* (London: Verso, 1976).

Ferdinand de Saussure, *Course in General Linguistics* (New York: Columbia University Press, 2011).

Thomas A. Schwandt, "Three Epistemological Stances for Qualitative Inquiry: Interpretativism, Hermeneutics and Social Constructionism," in *The Landscape of Qualitative Research: Theories and Issues*, edited by N. Denzin and Y. Lincoln, 292–331 (Thousands Oaks, CA: Sage).

John R. Searle, "Social Ontology: Some Basic Principles," *Anthropological Theory* 6 (2006): 12–129.

Karen Sermon et al., "Preimplantation Genetic Diagnosis," *The Lancet* 363 (2004): 1633–1644.

David Serre and Svante Pääbo, "Evidence for Gradients of Human Genetic Diversity Within and Among Continents," *Genome Research* 14 (2004): 1679–1685.

Steven Shapin, "History of Science and Its Sociological Reconstructions," *History of Science* 20 (1982): 157–211.

D. K. Simonton, "Francis Galton's Hereditary Genius: Its Place in the History and Psychology of Science," in *The Anatomy of Impact: What Makes the Great Works of Psychology Great*, edited by R. J. Sternberg, 3–18 (Washington, DC: American Psychological Association, 2003).

Christine E. Sleeter, "Advancing a White Discourse: A Response to Scheurich," in *Anti-Racist Scholarship: An Advocacy*, edited by James J. Scheurich, 42–45 (Albany, NY: SUNY Press, 2002).

Audrey Smedley and Brian Smedley, *Race in North America: Origins and Evolution of a Worldview* (Boulder, CO: Westview Press, 2012).

Lillian Smith, *Killers of the Dream* (New York: Norton, 1961).

Bonnie B. Spanier, *Im/partial Science: Gender Ideology in Molecular Biology* (Bloomington, IN: Indiana University Press, 1995).

Robert Sparrow, "A Not-So-New Eugenics: Harris and Savulescu on Human Enhancement," *Hasting Center Report* 41 (2011): 32–42.

Arthur K. Spears, "Negotiating Racism in the Academy," in *Racism in the Academy: The New Millennium*, edited by Audrey Smedley and Janis Faye Harrison, 146–154. The Commission on Race and Racism in Anthropology and the American Anthropological Association Report, 2012.

John H. Stanfield, II, "The 'Negro Problem' Within and Outside the Institutional Nexus of Pre-World I Sociology," *Phylon* 43 (1982): 187–201.

Nancy Stephan, *The Idea of Race in Science: Great Britain 1800–1960.* (Hamden, CT: Archon Books, 1982).

Anne Fausto-Sterling, "Refashioning Race: DNA and the Politics of Health Care," *Differences: Journal of Feminist Cultural Studies* 15 (2004): 1–37.

Jacqueline Stevens, "Racial Meanings and Scientific Methods: Changing Policies for NIH-Sponsored Publications Reporting Human Variation," *Journal of Health Politics, Policy and Law* 28 (2003): 1033–1087.

Lothrop Stoddard, *The Rising Tide of Color against White World Supremacy* (United Kingdom: Historical Review Press, [1921] 1985).

Hua Tang et al., "Genetic Structure, Self-Identified Race/Ethnicity, and Confounding in Case-Control Association Studies," *American Journal of Human Genetics* 76 (2005): 268–275.

Alan R. Templeton. "Human Races: A Genetic and Evolutionary Perspective," *American Anthropologist* 100 (1999): 632–650.

Eric C. Thompson, "The Problem of 'Race as a Social Construction,'" *Anthropology News* 47 (2006): 6–7.

Wenda R. Trevathan et al., *Evolutionary Medicine and Health* (New York: Oxford University Press, 2008).

Robert C. Tucker, *The Marx-Engels Reader*, 2nd edition (New York: W. W. Norton and Company, 1978).

William H. Tucker, *The Funding of Scientific Racism: Wickliffe Draper and the Pioneer Fund* (Urbana-Champaign, IL: University of Illinois Press, 2007).

U.S. Preventive Services Task Force, "Aspirin for the Primary Prevention of Cardiovascular Events: Recommendation and Rationale," *Annals of Internal Medicine* 136 (2002): 157–160.

U.S. Preventive Service Task Force on Prostate Screening, "Final Recommendation Statement," May 2012, http://www.uspreventiveservicestaskforce.org/Page/Document/RecommendationStatementFinal/prostate-cancer-screening#citation1.

J. Craig Venter, "The Richard Dimbleby Lecture 2007: Dr. J Craig Venter—A DNA-Driven World," *BBC One*, December 4, 2007, accessed March 31, 2011, http://www.edge.org/3rd_culture/venter.dimbleby07/venter.dimbleby07_index.html.

Priscilla Wald, "Blood and Stories: How Genomics is Rewriting Race, Medicine and Human History," *Patterns of Prejudice* 40 (2006): 303–333.

Danna L. Walker et al., "Inverting the Inverted Pyramid: A Conversation about the Use of Feminists Theories to Teach Journalism," *Feminist Teacher* 19 (2009): 177–194.

Alex Wellerstein, "Harry Laughlin's 'Model Eugenical Sterilization Law," *AlexWellerstein.com*, April 20, 2014, http://alexwellerstein.com/laughlin.

Frances E. White, "Africa on My Mind: Gender, Counter Discourse and African American Nationalism," *Journal of Women's History* 2 (1990): 73–97.

Cathy S. Widom and Linda M. Brzustowicz, "MAOA and the 'Cycle of Violence': Childhood Abuse and Neglect, MAOA Genotype, and Risk for Violent and Antisocial Behavior," *Biological Psychiatry* 60 (2006): 685–689.

David R. Williams, "Race, Socioeconomic Status, and Health: The Added Effects of Racism and Discrimination," *Annals of the New York Academy of Sciences* 896 (1999): 173–188.

Juan Williams, "Violence, Genes, and Prejudice," *Discover Magazine*, November 1, 1994, accessed June 8, 2015, http://discovermagazine.com/1994/nov/violencegenesand446.

Carter G. Wilson, *Racism: From Slavery to Advanced Capitalism* (Thousand Oaks, CA: 1996).

Tim Wise, "Famous Last Words: Exploring the Depth of Racist Conditioning," *TimWise.org*, April 24, 1999, http://www.timwise.org/1999/04/famous-last-words-exploring-the-depths-of-racist-conditioning.
Scott Woods, "5 Things No One Is Actually Saying About Ani DiFranco or Plantations," *Scott Woods Makes Lists blog*, January 3, 2014, http://wp.me/p40Hhl-1g.
Iris Young, "Gender as Seriality: Thinking about Women as a Social Collective," *Signs* 19 (1994): 713–738.
Michael Yudell et al., "Taking Race out of Human Genetics," *Science* 351 (2016): 564–565.
Slavoj Žižek, *How to Read Lacan* (New York: W. W. Norton & Company, 2007).
Slavoj Žižek, *The Plague of Fantasies* (London: Verso, 1997).
Slavoj Žižek, *The Sublime Object of Ideology* (London; New York: Verso, 1989).
Tukufu Zuberi and Eduardo Bonilla-Silva, eds., *White Logic, White Methods: Racism and Methodology* (Lanham, MD: Rowman & Littlefield, 2008).

Index

About the Author

Johnny E. Williams is associate professor of sociology at Trinity College. His research investigates how culture (i.e., shared beliefs, values, and meaning systems) maintains and challenges social order. He is the author of *African-American Religion and the Civil Rights Movement in Arkansas*.

www.ingramcontent.com/pod-product-compliance
Ingram Content Group UK Ltd.
Pitfield, Milton Keynes, MK11 3LW, UK
UKHW041104150926
475561UK00001BA/32